Engaging Eager and Reluctant Learners

Also by the Authors

Imagine, Inquire, and Create: A STEM-Inspired Approach to Cross-Curricular Teaching, 2nd Edition, Kindle Edition (2015)

Teaching Math, Science, and Technology in Schools Today: Guidelines for Engaging Both Eager and Reluctant Learners 2nd Edition (2014)

Demystify Math, Science, and Technology: Creativity, Innovation, and Problem-Solving, 2nd Edition (2013)

Tomorrow's Innovators: Essential Skills for a Changing World, Kindle Edition (2012)

Activating Assessment for All Students: Differentiated Instruction and Information Methods in Math and Science 2nd Edition, Kindle Edition (2012)

Shaping the Future with Math, Science, and Technology: Solutions and Lesson Plans to Prepare Tomorrows Innovators (2011)

Demystify Math, Science, and Technology: Creativity, Innovation, and Problem-Solving (2010)

Activating Assessment for All Students: Innovative Activities, Lesson Plans, and Informative Assessment (2009)

Helping Students Who Struggle with Math and Science: A Collaborative Approach for Elementary and Middle Schools (2008)

Engaging Eager and Reluctant Learners

STEM Learning in Action

Dennis Adams and Mary Hamm

ROWMAN & LITTLEFIELD
Lanham • Boulder • New York • London

Published by Rowman & Littlefield
A wholly owned subsidiary of The Rowman & Littlefield Publishing Group, Inc.
4501 Forbes Boulevard, Suite 200, Lanham, Maryland 20706
www.rowman.com

Unit A, Whitacre Mews, 26–34 Stannary Street, London SE11 4AB

© 2017 by Dennis Adams and Mary Hamm

British Library Cataloguing in Publication Information Available

Library of Congress Cataloging-in-Publication Data Available

ISBN: 978-1-4758-3445-1 (cloth : alk. paper)
ISBN: 978-1-4758-3446-8 (paper : alk. paper)
ISBN: 978-1-4758-3447-5 (electronic)

♾️™ The paper used in this publication meets the minimum requirements of American National Standard for Information Sciences—Permanence of Paper for Printed Library Materials, ANSI/NISO Z39.48-1992.

Printed in the United States of America

Contents

Preface vii

1 Basic STEM Concepts: Providing Different Avenues to
Understanding 1

2 STEM Instruction 17

3 Science and Mathematics: The Power of Inquiry and
Problem Solving 43

4 Engineering and Technology: Imaginatively Integrating
STEM Subjects in the Classroom 67

5 STEM: Past, Present, and Future 107

6 Evaluating STEM Learning: *In*formative Assessment,
Lesson Plans, and Activities 129

About the Authors 159

Preface

Engaging Eager and Reluctant Learners:STEM Learning in Action! builds on the interdisciplinary nature of the STEM disciplines. It encourages building on integrated possibilities of these subjects and focusing on students' learning needs, strengths, interests, and individual profiles. The book is designed for teachers who want ideas, activities, and lesson plans for an active participatory classroom.

We need to recognize the fact that the STEM subjects are frequently misunderstood and the risks difficult to analyze. The lack of STEM understanding has been cited as having a negative effect on everything, from the economy to national security. Technology merits special attention because it stands out in the popular imagination as wondrous, invasive, and unnerving.

In the classroom, it is usually best for the STEM subjects to come together and connect with real-world problems. When this happens, the result can be the opening of learning paths that reach across the curriculum. This may allow teachers and students to work with some of the big interdisciplinary ideas and questions. For example: "How might science and technology make us smarter?" "How can you avoid letting the Internet distract you from real-life experiences?"

This book builds on the social nature of learning to provide some practical suggestions for reaching every student in the classroom. We explore multiple intelligence theory because it is sometimes viewed as one of the foundational building blocks of STEM instruction. Also, many of the concepts and methods found here follow a pattern suggested by cognitive psychology and brain-compatible research.

Engaging Eager and Reluctant Learners: STEM Learning in Action! suggests careful consideration of students' learning tendencies and provides different avenues for learning and assessment. Also, for teachers and students to be enthusiastic and motivated requires a combination of teaching and assessment strategies that have been shown to have positive effects on student learning.

In the classrooms of experienced teachers, assessment tools and collaborative learning experiences are approached as natural partners. They know that motivation to learn often stems from the connection between authentic assessment and teaching. So, we give special attention to performance assessment and building portfolios.

We believe that the best way to motivate learners is to focus on individual interests, cooperative inquiry, and problem solving. Basic skills matter. The same can be said for establishing roots in what has worked in the past—a proven way to move forward toward new approaches and future uncertainties.

For all students to reach their full potential often requires the modification of content, process, and product for both individuals or small clusters of students. But no matter how much you differentiate or individualize it, everyone on a learning team has to work together to gain an understanding of the basic concepts and skills being taught. To paraphrase Vygotsky, *what students can do together today, they can do alone tomorrow.*

How students characteristically deal with adversity says a lot about their future imaginative achievements. Few things go perfectly. So, it is best to prepare for a few unpleasant surprises. In addition, it is important for everyone to understand that creativity, new ideas, and everyday life depends on overcoming setbacks and disappointments. We also have to confront the undeniable fact that people in the future will have to live with the decisions we make today.

Exemplary teachers are efficient, inventive, pragmatic, and forward-looking. They may take surprisingly different approaches to instruction, but a common question they ask is *does it work?* If something new is interesting, try it. If it works, keep using it; if it doesn't, push it aside and try something else. The motto is: *do your best today; do better tomorrow.*

Hopefully, this book opens some unique doors to instruction in a way that helps teachers provide more opportunities for learners to construct their own knowledge. The basic idea is to provide teachers with some new ideas and methods so that they can go about organizing science, technology, engineering, and math instruction in a more connected way.

Experienced teachers know that individual learners need to approach learning in different ways. Like more than a few teachers, some learners simply don't like certain subjects, and others don't think they can be successful. At any level, it seems that poor attitude and poor achievement can amplify each other.

Developing positive attitudes toward science, technology, engineering, and mathematics goes hand in hand with developing competency. The biggest challenge, many teachers say, is trying to engage students and persuade them that all of the STEM subjects matter. With math and science, for example, more than a few students follow motto of *bad attitude, how to get it, and how to keep it.*

Instruction is made more difficult when teachers are not prepared to teach a subject. On top of that, some teachers are more than a little afraid of one or more of the STEM subjects. The good news is that many educators are becoming more familiar with collaborative hands-on learning. As a consequence, approaches that emphasize teamwork provide a relatively easy way for them to approach subjects that might be viewed as difficult.

Good teachers know that involving students in active, participatory, and connected learning is the best approach available for individual students to come to terms with the subjects they would rather avoid. As you go about amplifying learning, remember that academic success often comes from excursions that are made from a secure and orderly base.

Engaging Eager and Reluctant Learners: STEM Learning in Action! is written in a teacher-friendly style intended to help with classroom organization and lesson planning. Although it is intended for individual teachers, the book is organized in a way that is convenient for school districts that are doing in-service work. We focus on topics that are especially useful for elementary and middle school teachers, assisting them as they invite learners to inquire, discover concepts, and collaboratively explore interlinking concepts.

We may all be a little ambivalent about digital technology. But whether its influence is good, bad, or indifferent, tech products have thoroughly invaded and transformed our lives over the last three decades. We are all trying to understand the rapidly shifting implications so that some control can be exerted over the technological products of science, engineering, and math.

As technology lessons are developed, it is important to include instruction on distinguishing between what is and what isn't useful information. Also, it is up to teachers to shine some light on some of the mysterious processes involved, while enhancing student understanding of the STEM subjects. By demystifying practice in these areas of study, teachers can help all of their students gain a better understanding of science, technology, engineering, and mathematics. Hopefully, this can all be done without projecting our worries onto their growing visions of the future.

In addition to looking just over the horizon, it is hoped that *Engaging Eager and Reluctant Learners: STEM Learning in Action!* will deepen the collective conversation and help teachers reverse the steady erosion of STEM skills in the general population.

Chapter One

Basic STEM Concepts

Providing Different Avenues to Understanding

Galileo is famous not just because he saw how the stars move, he's famous because he insisted we see for ourselves how the world works, share what we see and shape our society accordingly.

—Alice Dreger

This book provides many adaptive STEM approaches, strategies, and lesson plans that are intended to help students understand how the world and the universe work. The basic idea is to creatively integrate elements of science, technology, engineering, and mathematics instruction in ways that result in shared narratives. Creativity and innovative problem solving are core concerns.

Building on the social nature of learning is central to connecting and amplifying a web of motivational, attitudinal, and academic abilities. At school and at work, diverse groups are more likely to shine. When teams are composed of individuals who are too similar to each other, they are more likely to ignore vital information, become overconfident, and make poor decisions (Johnson, 2014).

We live in a world defined by smartphones, Android or Apple, fast networks, and cloud computing. Still, critical thinking, problem solving, and a solid knowledge base are skills that are more important than ever. The best STEM instruction recognizes these facts and is open-ended and inquiry based.

STEM is more of a philosophy than a specific program. Just as there are multiple paths to competency, there are different ideas and principles for guiding instruction. The foundational building blocks include cognitive science, new media, and culturally fair assessment. One of the goals is to help students understand that the solution to problems should be based on how things are, not how we would like them to be.

1

As schools are adapting to an increasingly diverse group of learners, it is important to design science, technology, engineering, and mathematics instruction in ways that accommodate all learners. Inclusion, connection, and collaborative learning can be arranged in ways that ensure that fewer children will be left behind and high achievers are challenged.

Educators have long known that schools are places where both teachers and students can learn from each other. Although integrating the STEM subjects is a more recent development, it is an easy fit for teachers who are used to arranging small group instruction. Sometimes students work in pairs to solve problems. At other times, implementing STEM-influenced learning is as simple as reemphasizing a question.

PAST SUCCESSES AND FUTURE ACHIEVEMENTS

Assessing student work before, during, and after instruction can improve tomorrow's lessons. Teacher observation has always been important. Also, informative assessment can take place before at any stage of a lesson. The basic idea is to help teachers identify the needs and interests of their students. The next step is adapting instructional plans in a way that maximizes the success of all learners.

Quality teaching involves tailoring instruction to meet the needs of students who have a wide range of needs and academic abilities. One size doesn't fit all. But individuals within small groups can use different processes and develop unique products on the way to achieving similar conceptual understanding.

How a subject is taught and what students learn are important elements of any lesson evaluation. The same can be said for student and teacher evaluation of how well the group worked together. Sample question: Can you name one thing that went well and two things that could make the group or lesson better?

Although teaching and learning are the emphases here, it is important to keep in mind that improving political and school culture matters as much as student motivation and teacher effectiveness. Also, political and educational leaderships have to make sure that the appropriate resources for the schools are in place.

Clearly, good administrators, a positive school culture, and imaginative development strategies serve as the foundation for improving individual learning in the classroom.

When teachers are looking for factors that get in the way of student achievement, it is possible to create classroom routines that support STEM learning. It may be impractical to fully individualize instruction, but the pace of learning can be easily varied. In addition, learning centers can be designed to accommodate different learning styles and abilities in ways that allow small clusters of students to work together on shared interests and needs.

STEM instruction involves building mixed-ability group instruction around the idea that individual students (or clusters of students) learn in unique ways and at varying levels of difficulty. Assessment can follow a similar path. Along with teacher observation, formative assessment can serve as a powerful to adjust lessons and future grouping decisions.

A good way for teachers to integrate the STEM subjects and extend their reach across the curriculum is to base group assignments on what is known about the interests and aptitudes of the students involved. In addition, there are times when students may need multiple chances to demonstrate mastery.

In a STEM-influenced classroom, you will find students doing more thinking for themselves and more work with peers. Self-evaluation and a gentle kind of peer assessment are parts of the process. With STEM instruction, small collaborative groups within the class are sometimes working at different levels of complexity and at different rates. Lessons can be based on students' interest in a subject, readiness to learn a concept, or their preferred path to comprehension.

Instruction works out better when concepts are taught in context and related to prior knowledge. STEM lessons can also provide multiple paths to understanding and expressing what has been learned. The process involves having learners construct meaning by working with peers to explore issues, problems, and solutions. In this way, STEM is different from individualized instruction that attends to specific student needs as STEM concentrates on the needs of clusters of students.

To paraphrase Vygotsky, *what children can do together today, they can do alone tomorrow.*

THE EVOLUTION OF IDEAS ABOUT THINKING AND LEARNING

A number of new theories and research findings have come from psychology, cognitive science, and related research about how the brain functions. Many of these ideas fit in nicely with the concept of STEM. We know many things about teaching, learning, and how the mind works that we didn't know about even a few decades ago. For the last decade or so, researchers have been trying to understand the mind's capabilities and figure out how the results might be applied to learning.

Do experts in the STEM subjects always need experimental evidence to be considered credible? Maybe, but at the cutting edge of knowledge the answer isn't always that clear. Some scientists, for example, suggest that good science doesn't always have to be tied to empirical evidence (Frank & Gleiser, 2015).

In any field a creative and compelling argument sometimes has to be considered useful before the research can catch up with it. Also, in any professional field, there are times when strongly supported evidence is walled off from everyday decisions. The various standard projects have helped by building on the research to inform practice. But it doesn't always help when we give more and more of our lives over to algorithms.

Although connecting the research to actual classroom practice is sometimes a problem, the gap is being narrowed. STEM learning not only builds on new ideas gleaned from brain research and goes on to suggest using a balance of visual, auditory, oral, and written materials to match the preferences of different kinds of learners.

Reading science fiction can provide a critical reflection on the present. But it's not so useful when it comes to predicting the future. Many of the jobs our students will be called on to do in the future don't even exist yet. So, it would be a mistake to pay too much attention to narrow vocational skills because there is a risk that students may see those skills rapidly going out of date.

Back in the early 1980s many of us spent a lot of time learning the BASIC programming language. Was it all that helpful in the long run? So more important skills for today and tomorrow: creativity, problem solving, risk-taking, and teamwork. These competencies and concerns are bound to be key skills for any future scenario.

The field of STEM education is constantly growing and changing. We now have a better understanding of both the problems and the possibilities associated with the teaching of science, technology, engineering, and math. Several libraries could be filled with books, journals, research papers, and projects that relate to the expanding educational knowledge base. Still, a synthesis of the most important recent findings and how they relate to the classroom would certainly help.

Until well past the midpoint of the twentieth century, the theoretical ideas about learning were dominated by a behaviorist view of rewards and punishment. Over the last few decades, the cognitive perspective has largely taken its place. Cognitive science provides ways for thinking about how the mind works and how knowledge is acquired and represented in the memory system. Developments in neuroscience brain research have further extended the field; both support the idea that differentiation can inspire a student's best effort.

Cognitive science, multiple intelligence (MI) theory, individualized learning, and constructivism are at least indirectly related. Constructivist educators emphasize teaching students to classify, analyze, predict, create, and problem solve. Student ability to learn new ideas is viewed as having a lot to do with the information an individual has prior to instruction.

Facts can be important building blocks, but knowledgeable teachers emphasize actively building new structures on prior knowledge. This constructivist approach is based on practices that encourage hands-on experiences that encourage students to build on prior experiences to discover concepts and models that help them understand the world around them.

Good teachers know that students from different cultural and economic backgrounds may need extra inspiration and multiple approaches for learning science, technology, engineering, and math. With technology, for example, it's often surprising to see what clever people actually do with new ideas and inventions. It is little wonder that applications frequently come along that the inventors or developers haven't even dreamed of.

STEM instruction can help students deal with the new and can even help develop future innovators fill in the gaps between what wonderful things they imagine and the sometimes sad first results. Revision is one of the keys to success. When it comes to instruction, teachers can build on just about any theoretical model by changing the pace, level, and type of instruction to meet the learning needs of students from different strengths, weaknesses, preferences, and needs.

In a STEM-influenced classroom, carefully designed student-centered learning and thinking about how you are teaching is a good way to ensure that students with diverse abilities and interests are well served. Critique, evaluation, and analysis all help. It also helps if you take the time to study the research on effective instruction and try to apply the relevant theory and facts in the classroom.

MULTIPLE ENTRY POINTS TO SCIENCE, TECHNOLOGY, ENGINEERING, AND MATH KNOWLEDGE

Many questions remain, but no one doubts that today's students have an increasingly diverse range of needs, abilities, and interests. To make science, technology, engineering, and mathematics more accessible to today's students means respecting multiple ways of making meaning.

The brain has a multiplicity of functions and voices that speak independently and distinctly for different individuals. Clearly, learning has a lot to do with finding your own strengths and preferred paths to knowledge.

An enriched environment for one student is not necessarily enriched for another. In the STEM classroom, teachers try to maximize each student's learning capacity with a systematic approach to academically diverse learners. The emphasis is on student learning needs and using small groups to enhance each student's learning capacity.

In the STEM classroom, teachers are aware of who their students are and how student differences relate to what is being taught. By having the flexibility to differentiate or adapt student work, it is possible to increase the possibility that each student will learn.

Elements that Guide STEM Learning

Five ways teachers can differentiate or modify instruction:

The *content* that teachers teach and how students have access to information is an important way for teachers to differentiate instruction. Student readiness is the current knowledge understanding and skill level of a student. Readiness does not mean student ability; rather, it reflects what a student knows, understands, and is able to do.

Interest is another way to individualize learning. Topics students enjoy learning about, thinking about, and doing provide a motivating link. Successful teachers incorporate required content to students' interests to engage the learners. This helps students connect with new information by making it appealing, relevant, and worthwhile.

A student's *learning profile* is influenced by an individual's preferred learning style, "intelligence" preference, academic interests, and cultural background (Gardner, 2006). By tapping into a student's learning profile, teachers can extend the ways students learn best.

A STEM learning environment enables teachers and students to work in ways that benefit each student and the class as a whole. A flexible environment allows students to make decisions about how to make the classroom surroundings work. This gives students a feeling of ownership and a sense of responsibility. Students of any age can work successfully as long as they know what's expected and are held to high standards of performance.

> *To individualize instruction, you need to clarify* the content *(what you want students to know and be able to do),* the process *(how students are going to go about learning the content), and* the product *(how they will show what they know).*
>
> —Amy Benjamin

IMPORTANT PRINCIPLES OF STEM INSTRUCTION

There are several key principles that describe a STEM classroom. A few of them are defined here:

1. *A high-quality engaging curriculum* is the primary principle. A teacher's first job is to guarantee the curriculum is consistent, inviting, important, and thoughtful.

2. Students' work should be appealing, inviting, thought provoking, and stimulating. *Every student should find his or her work interesting and powerful.*

3. Teachers should try to *assign challenging tasks* that are a little too difficult for the student. Be sure there is a support system to assist students' success at a level they never thought possible.

4. *Use adjustable grouping.* It is important to plan times for groups of students to work together—and times for students to work independently. Provide teacher-choice and student-choice groups.

5. *Assessment is an ongoing process.* Preassessment determines students' knowledge and skills based on students' needs. Then teachers can individualize instruction to match the needs of each student. When it's time for final assessments, it's helpful to plan several assessment strategies—for example, a quiz and a project.

6. *At least some grades should be based on growth.* A struggling student who persists and doesn't see progress will likely become frustrated if grade-level benchmarks remain out of reach and growth doesn't seem to count. It is the teacher's job to support the student by making sure that (one way or another) they master the concepts required.

When it comes to planning activities or assignments, we often use a tiered approach. The basic idea is to have a wide range of students learn the concept being taught, but can reach competency in unique ways. The first step is to identify the key skills and concepts that everyone must understand.

All students in a class cover the same topic, but the teacher varies materials based on students' aptitudes and interests. For example, one learning group can write (in an expository way) about how science builds on empirically tested laws—while another group documents how specific scientific evidence from the past is used to formulate questions or build a hypothesis.

As one group writes a report, another might use video camera, webcam, iPad, or smartphone to construct a six-, eight-, or ten-minute video to share online.

We use YouTube because this free site simplifies the process of uploading or downloading videos—and they can reach a large audience. (It's also fine if the more print oriented group also wants to construct a video [in their style] for the Internet.)

Innovation requires investing in research. If you want to look into the day-to-day life of scientists solving problems and doing imaginative work, go to Research2Reality (https://research2reality.com). This site has two-minute videos of dozens of scientists doing basic, curiosity-driven research work. They briefly present to an easy-to-comprehend picture of what they are doing to help build the foundations of future innovation.

Over the last few decades, researchers have suggested that since the human brain is "wired" in different ways, it is important for teachers to realize that students learn and create in different ways. Although it is often best to teach to a student's strength, we know that providing young people with deep learning experiences in different domains can enrich their "intelligence" in specific areas.

Howard Gardner and Robert Sternberg have contributed to the awareness that students exhibit different intelligence preferences. Sternberg suggests three intelligence preferences: analytic (schoolhouse intelligence), creative (imaginative intelligence), and practical (contextual, street-smart intelligence). Gardner suggests at least eight.

MULTIPLE INTELLIGENCES, ONE OF THE BUILDING BLOCKS OF STEM INSTRUCTION

The biggest mistake of past centuries in teaching has been to treat all children as if they are variants of the same individual, and thus to feel justified in teaching them the same subjects in the same ways.

—Howard Gardner

Howard Gardner's framework for multiple entry points to knowledge has had a powerful influence on differentiated learning and the content standards. Both the math and the science standards recognize alternate paths for learning. There are many differences, but each set of content standards is built on a belief in the uniqueness of each child and the view that this can be fused with a commitment achieving worthwhile goals.

Being able to base at least some math and science instruction on a student's preferred way of learning has proven to be especially helpful in teaching struggling students. It's not just academics. Physical activity can have a positive effect on performance tasks requiring concentration.

When Gardener was writing *Frames of Mind* (outlining MI theory), he was critical of how the field of psychology had traditionally viewed of intelligence. So, he set out to stir up some controversy. He succeeded.

An unexpected result of his writing about multiple intelligences was the enthusiastic response within the educational community. Teachers showed unexpected interest in exploring MI theory and putting some activities based on that theory into practice. In fact, lessons built around Gardner's concept of multiple intelligences proved to be particularly helpful in meeting some of the challenges of heterogeneous grouping and an increasingly diverse student body.

Multiple Intelligences

1. Linguistic intelligence: the capacity to use language to express ideas, excite, convince, and convey information; includes speaking, writing, and reading.
2. Logical-mathematical intelligence: the ability to explore patterns and relationships by manipulating objects or symbols in an orderly manner.
3. Musical intelligence: the capacity to think in music, the ability to perform, compose, or enjoy a musical piece; includes rhythm, beat, tune, melody, and singing.
4. Spatial intelligence: the ability to understand and mentally manipulate a form or object in a visual or spatial display; includes maps, drawings, and media.
5. Bodily kinesthetic intelligence: the ability to use motor skills in sports, performing arts or art productions (particularly dance or acting).
6. Interpersonal intelligence: the ability to work in groups—interacting, sharing, leading, following, and reaching out to others.
7. Intrapersonal intelligence: the ability to understand one's inner feelings, dreams, and ideas; involves introspection, meditation, reflection, and self-assessment.
8. Naturalist intelligence: the ability to discriminate among living things (plants, animals) as well as sensitivity to the natural world.

Most academic work and vocations involve several of these "intelligences." Gardner started with seven, added an eighth, and has tentatively suggested a ninth. In the assessment chapter, we view multiple intelligences a bit differently and list *existential* intelligence as number nine; it involves the ability to think about questions or phenomena beyond sensory data.

Within any variation of the MI framework, intelligence might be defined as the ability to solve problems, generate new problems, and do things that are valued within one's own culture. MI theory suggests that these eight "intelligences" work together in complex ways. Most people can develop an adequate level of competency in all of them. And there are many ways to be "intelligent" within each category. Gardener's "intelligences" may have been central to the twentieth century, but what about the twenty-first century?

It is possible to take issue with MI theory on other points, like not fully addressing spiritual and artistic modes of thought. But there is general agreement on a central point: *intelligence is not a single capacity that every human being possesses to a greater or lesser extent.*

No matter how you prefer to explain it, there *are* multiple paths to competency in math and science. As a consequence, it makes instructional sense to differentiate instruction in a way that builds on different ways of knowing and understanding. For students to learn personally, across learning profiles, they need rich materials that allow room for varying responses.

A certain level of differentiation is hard to avoid. Even if we give students the same materials, each student's background and strengths will result in somewhat different experiences and results.

Activities that Reflect Multiple Intelligence Theory

1. Upper elementary and middle school students can comprehend MI theory. Why not explain it to them and have them do some activities to remember it?

We like having students work with a partner. Here are some possibilities:

Linguistic intelligence	*Musical intelligence*
Write an article	Sing a rap song
Develop a newscast	Give a musical presentation
Make a plan, describe a procedure	Explain music similarities
Write a letter	Make and demonstrate a musical
Conduct an interview	instrument
Write a play	Demonstrate rhythmic patterns
Interpret a text or piece of writing	
Logical-mathematical intelligence	*Spatial intelligence*
Design and conduct an experiment	Illustrate, draw, paint, sketch
Describe patterns	Create a slide show, videotape,
Make up analogies to explain ...	chart, map, or graph
Solve a problem	Create a piece of art
Bodily kinesthetic intelligence	*Interpersonal intelligence*
Use creative movement	Conduct a meeting
Design task or puzzle cards	Participate in a service project
Build or construct something	Teach someone
Bring hands on materials to demonstrate	Use technology to explain
Use the body to persuade, console or	Advise a friend or fictional
support others	character
Naturalist intelligence	*Intrapersonal intelligence*
Prepare an observation notebook	Write a journal entry
Describe changes in the environment	Describe one of your values
Care for pets, wildlife, gardens, or parks	Assess your work
Use binoculars, telescopes, or microscopes	Set and pursue a goal
Photograph natural objects	Reflect on or act out emotions

2. Encourage various learning styles:

Mastery style learner	Concrete learner, step-by-step process, learns sequentially
Understanding style learner	Focuses on ideas and abstractions learns through a process of questioning
Self-expressive style learner	Looks for images, uses feelings and emotions
Interpersonal style learner	Focuses on the concrete, prefers to learn socially, judges learning in terms of its potential use in helping others

3. Build on students' interests.When students do research either individually or with a group, allow them to choose a project that appeals to them. Students should also choose the best way for communicating their understanding of the topic. In this way, students discover more about their interests, concerns, their learning styles, and intelligences.
4. Plan interesting lessons. There are many ways to plan interesting lessons.

      (Lesson plan ideas presented here are influenced by ideas as diverse as those of John Goodlad, Madeline Hunter, and Howard Gardner.)

LESSON PLANNING

1. Set the tone of the lesson. Focus on students' attention. Relate the lesson to what students have done before. Stimulate interest.
2. Present the objectives and purpose of the lesson. What are students supposed to learn? Why is it important?
3. Provide background information: what information is available? Resources such as books, journals, videos, pictures, maps, charts, teacher lecture, class discussion, or seat work should be listed.
4. Define procedures. What are students supposed to do? This includes examples and demonstrations as well as working directions.
5. Monitor students' understanding. During the lesson, the teacher should check students' understanding and adjust the lesson if necessary. Teachers should invite questions and ask for clarification. A continuous feedback process should be in place.
6. Provide guided practice experiences. Students should have a chance to use the new knowledge presented under direct teacher supervision.

7. It is equally important that students get opportunities for independent practice where students can use their new knowledge and skills.
8. Evaluate and assess students' work to show that students have demonstrated an understanding of significant concepts.

Activity 1: Introduce Graphic Organizers

Graphic organizers help students retain semantic information. Mind mapping or webbing illustrates a main idea and supporting details. To make a mind map, write an idea or concept in the middle of a sheet of paper. Draw a circle around it. Then draw a line from the circle. Write a word or phrase to describe the idea or concept. Draw other lines coming from the circle in similar manner. Then have students draw pictures or symbols to represent their descriptions.

Musical	Singing songs about neurons, tapping out rhythms to the song "Because I Have a Brain"
Naturalist	Describing changes in your brain environment, illustrating a dendrite connection
Interpersonal	Participate in (act out) a group signal neuron transmission observing/recording
Intrapersonal	Reflecting on being a neuron, keeping a journal of how the brain works
Mathematical/logical	Calculating neuron connections

Assessment

Each group will write a reflection on the activity. Journal reflections should tell what they learned about neurons and how that helps them understand how the brain works. Encourage students to organize their work and put it in a portfolio.

Student interests and strengths have to be considered. So, MI theory and learning style have roles to play in STEM instruction. Even during direct instruction, lessons can be modified to meet the needs, interests, and aptitudes of individuals and small clusters of students.

SORTING TRUTH FROM FICTION

The STEM subjects provide powerful ways of sorting the truth from fiction and transforming human existence. In science, for example, important findings are evaluated and other investigators often try to replicate discoveries and reinterpret meaning.

STEM instruction can be an organized and flexible platform for adjusting teaching and learning to meet students where they are; at the same time, learners can be encouraged to help them accomplish more academically. Since many of the jobs students will have don't exist yet, this isn't the best time to narrow the curriculum to basic subskills.

Self-reliant, imaginative, and motivated learners are key to success in the future. Online activities may help with agility and multitasking, but the end result may be an inability to explore narrative in ways that place ideas, people, and events in a wider context.

The human brain adapts to the environment, and today's environment is changing in unprecedented ways. One result is a loss of attention span that interferes with the ability to think deeply about complex issues. Baroness Susan Greenfield suggests that the Internet is rewiring students' brains in negative ways.

Are sites like YouTube and Twitter contributing to the fragmentation of our culture? E-mail, Twitter, Facebook, and Instagram are but three examples of modern media functions that change the social universe while changing how we think.

These days, children are more likely to own an iPhone than a book. With all the time young people spend online, it's little wonder that so many of them have trouble concentrating for a long period of time. It can also lead to the development of beliefs that run contrary to the facts (Greenfield, 2015).

Politicians sometimes use sketchy science to make political points. Big money does the same thing to increase profits. Many claims of technological breakthroughs quickly fade away. Early in the personal computer age, companies like Microsoft would come up with false reports of future products ("vaporware") to throw smaller companies off an innovative track.

Is it real science or convenient opinion? For example, when a highly motivated political group sees a scientific finding that doesn't fit their script, they all too often try to twist a factual story into a false one that serves their self-serving opinion. So, it is important that students learn how to notice when politics, commercial interests, or public relations distort science to fit a predetermined position.

Digital technology can help in the classroom, but it is a poor substitute for personal interaction. By the middle grades, many schools have students doing some of their work with computers. Outside of school children of all ages already spend too much time using electronic media. Putting some kind of a limit on screen time would certainly help.

Although television is still the dominant medium, computers, tablets, and smartphones are increasingly cutting into TV time. Will the increasingly frantic pace of the knowledge explosion make for a better world? Maybe. Peter

Nowak, for example, suggests that better side of our nature will be amplified up by science and technological advances (Nowak, 2015).

Anyone who wants to figure out how far a unique combination of the STEM subjects might take us might in the future will find some helpful hints in the last century. German historian Philipp Blom, for example, suggests that in the twentieth century the most important events were not the two world wars. *Neither of the world wars was the defining event of the twentieth century. Both were almost side effects of the same vast revolution of modern technology and the Enlightenment* (Blom, 2015).

SUMMARY, CONCLUSION, AND LOOKING AHEAD

Now, more than ever, the STEM subjects matter at school and in life. Good instruction in science, technology, engineering, and mathematics deals pays close attention to big questions. It is inquiry based and blurs the line between the four disciplines. In addition, ideas, activities, and projects can be integrated into lessons from language arts to social studies. Since elementary teachers tend to be generalists, they are uniquely qualified to teach such interdisciplinary lessons.

Teachers who themselves are not well educated are not going to educate their students to the level they need. What to look for: understanding the characteristics of effective instruction has proven to be a solid asset for teachers trying to reach students who are performing at varying levels.

The key is figuring out how to help young people understand the STEM subjects and encouraging them to be bold enough to collaboratively turn ideas and concepts on their head. Experienced teachers also realize that there is no conflict between nurturing individuals and promoting collaborative learning.

While building on group cooperation, teachers can provide different paths for learning science, technology, engineering, and math. By having the opportunity to collaboratively explore ideas, even unmotivated students tend to respond to appropriate challenges and enjoy learning about science and math. Flexible grouping and pacing, tiered assignments, and performance assessment can inject fresh energy into lessons.

Since students don't all learn at the same rate, it is important to consider the pacing of instruction when figuring out the adaptivity options. Good STEM instruction is responsive to specific individual and small group needs, as well as class performance as a whole.

Successful classrooms are full of energy, excitement, and the possibility of teaching all students no matter what their preferred learning modality. Having a positive attitude has a lot to do with developing individual aptitudes and teamwork skills.

The best STEM instruction is open-ended and inquiry based. With all the possibilities, teachers are coming to view it as an important ally in meeting the needs of students with increasingly diverse levels of prior knowledge, interests, and cultural backgrounds.

STEM-influenced instruction builds on the use of curiosity in a disciplined way. Students also learn that the interaction with other learners creates more learning, innovation, and creativity. It may also help students realize that education is a good way to light up their future.

Nothing in personal or professional life is static. Things are bound to change, philosophies will flip, and new approaches will emerge. None of us can escape the march of history. But we can prepare ourselves to deal with incessant change and get ready to reinvent the realities of this world. Although foreseeing future events is impossible, getting ready for the unexpected certainly helps. And remember, to get hints about what's just around the corner look around today. To paraphrase William Gibson, *some pieces of the future are already here, they're just not evenly distributed.*

REFERENCES

Benjamin, A. (2003). *Differentiated instruction: A guide for elementary teachers.* Larchmont, NY: Eye on Education.

Blom, P. (2015). *Fracture: Life and culture in the west, 1918–1939.* New York, NY: Basic Books.

Gardner, H. (2006). *Multiple intelligences: New horizons.* New York, NY: Basic Books.

Greenfield, S. (2015). *Mind change: How digital technologies are leaving their mark on our brains.* London, UK: Rider Books [Random House Group].

Nowak, P. (2015). *Humans 3.0: The upgrading of the species.* Guilford, CT: Lyons Press. Globe/Pequot [Roman & Littlefield].

RESOURCES

Lesson Planning

A Lesson Plan Format for Direct Instruction

Topic: *Grade Level:*

Objective:

Theme and/or Motivation:

Materials:
a.
b.
c.

Launching the Lesson:

Whole-Class Teacher Instruction:
List the concepts, definitions, and processes to be used in instruction.
List the directions for activities and the examples you will use.

Class/Group/Individual Activities:

Include several harder and several simpler problems that can help students at all levels
    of competency learn the same concept.
How are you going to provide for different interests, needs, and aptitudes?

Summarize:
How will you decide whether students have learned what you wanted?

Lesson Plan Outline for Group Investigations

Topic of Lesson: *Grade Level:*

What do you want students to learn?

Why are the concepts important?

What background information do students need before starting?

Organization and Procedures:

List the materials needed:
a.
b.
c.

How are you going to get the students involved?

Lesson development, questions, and desired product:

Small group options:

Gearing up (if the lesson is too easy):

Gearing down (if the lesson is too hard):

Assessment (observations, products produced, portfolio entry, etc.):

Chapter Two

STEM Instruction

STEM education is an interdisciplinary approach to learning which removes the traditional barriers separating the four disciplines of science, technology, engineering, and mathematics, and integrates them into the real world, rigorous and relevant learning experiences for students.

—Vasquez et al.

The typical STEM lesson combines two or more disciplines. Teaching and learning often reflect the fact that knowledge can be generated by observation, experimentation, using technology, and building models. Children are natural scientists, technology users, engineers, and mathematicians; their enthusiasm for inquiry, design, and problem solving is evident.

Science instruction often serves as a base for generating interest and energy in the other STEM subjects. Also, learning activities are sometimes constructed in ways that reach across the entire curriculum. It's not just basic skills and underlying concepts. In traveling the path to deeper learning, it is often necessary to go beyond the agreed upon facts to deal with interesting ambiguities.

The terrain around any subject has areas that are characterized by vague and conflicting information. Discovery is not a neat and linear process. Sometimes, the questions matter even more than the answers (Firestein, 2012). Of course, the more students know, the better questions they can ask; the answers they come with frequently generate new questions.

When working along the thin line between the known and unknown, it is important to guard against preconceptions. On a personal level, both competency and wisdom can be generated by a combination of self-understanding and teamwork.

STEM instruction has as much to do with teaching the *processes* of a subject as it does with the teaching content. For example, the *methods* that scientists use to discover facts in the natural world often matter as much as the facts themselves. To get a real feel for how scientists, technologists, engineers, and mathematicians work, it is best to go beyond what they do to *figure out what they are trying to do*.

The methods that experts in various fields use are just as important as what they discover. For example, understanding the scientific method is a key intellectual tool with many applications. In the classroom, teachers can provide opportunities for students to explore, identify with, apply, interpret, and assess themselves. How well students come to understand and "own" the knowledge is shown when they apply knowledge to everyday situations.

Like many subjects, technology is an amplifier of what is going on in the classroom. Good teaching can be enhanced, but it won't turn bad lessons into good ones. The Internet, for example, can be helpful for projects and guided research, but when used improperly, it can be a distraction. Like other subjects, technology-based learning is enhanced when it is arranged in a way that helps students develop skills that are in line with those used by adults in everyday life.

In the future, less than 20 percent of our students will become experts in one the STEM subjects. But whatever they do, every student can profit from the intellectual tools that come with learning about these subjects. In addition, those who are not going to become scientists, mathematicians, or engineers must have some understanding of these fields in order to be responsible and productive citizens.

STEM INSTRUCTION AND "SCIENTIFIC LITERACY"

An important goal of STEM instruction is to open students' minds and help them develop the ability to solve everyday problems. Another objective is to expand students' perception and appreciation of the very nature of the world—including water, rocks, plants, animals, people, and other elements of the natural world around us. The basic idea is to enhance the curiosity, interest, and knowledge of learners as they explore the world in which they live.

"Scientific literacy" is a term that has been around for over fifty years. It is still a major instructional goal today. STEM literacy is newer and broader; it weaves together competency in science, technology, engineering, and mathematics. In addition, it can provide connections to a whole range of subjects in the curriculum of National Assessment Governing Board (National Research Council, 2012).

Understanding what it means to be scientifically literate, for example, requires that each individual truly understands the subject as well as the nature of science and how it was developed (Lederman, 2014). The following is a list of five ways of understanding STEM (NGSS Lead States, 2013). With STEM subjects:

- knowledge is based on evidence.
- knowledge can be improved with new evidence.
- models, laws, procedures, and theories help explain nature.
- the content acknowledges the order, consistency, and numbers in nature.
- exploring content addresses questions about the nature of the world and its problems.

Whatever the subject, students learn best when they experience things themselves. Observing, estimating, measuring, collecting, and classifying are some of the ways of learning that come naturally to most students. The same can be said for the ability to imagine possible futures and to choose among them (Wilson, 2014).

It helps if teachers become familiar with the standards and goals of the STEM subjects they are teaching. The next step is letting these subjects blend into just about every aspect of classroom life.

SUBJECT MATTER EXAMPLES

The past quarter century has seen many pressures to reinvent goals and effect changes in science, technology, engineering, and mathematics. At the same time, many changes are taking place in the culture and in instructional practices (to say nothing the way people learn, live, and work). In an effort to respond to these conditions, new approaches have been identified to update traditional concepts and principles.

Do changes in our approach to teaching represent a reinvention of goals or curricula? The science curriculum (biology, chemistry, physics, and earth and space science) is a good example. The most important change in science education reflects changes in the nature of the discipline itself.

Technology, engineering, and mathematics may be viewed as tools of science. And research in the sciences today is just as concerned with finding solutions to personal and social problems as it is on dealing with theories related to the natural world. When it comes to new ideas and innovation, *curiosity* is one of the key variables.

The processes involved matter even more than the facts of a subject. In science, for example, the method and processes are key ingredients. All the STEM subjects may provide new options.

How else are we going to face the issues of the 21st century on things like the environment, health, security, food, and energy?

—John Gibbons

The problems schools face today are more difficult and involved than ever before. New technology-influenced social norms are emerging. Today's national culture is defined by a global economy, an information era, and differing family structures.

The world has become a knowledge-intensive place that changes how we think about many things, including teaching and learning. Collaboration and participation have become key ingredients in making a difference in the lives of students.

Along with active learning, the National Science Education Standards and *STEM Lesson Essentials* suggest that subjects:

1. have greater depth and less superficial coverage.
2. focus on inquiry and problem solving.
3. emphasize skills and knowledge of the subject.
4. accommodate individual differences and creativity.
5. contain a common core of subject matter.
6. be closely coordinated with related subjects.
7. be attuned to personal relevance to a student (Brandt, 2000).

One of the changes from past curriculum designs is adding more depth and less superficial coverage to the content. Focusing in depth on a smaller number of skills and concepts leads to greater understanding and retention.

A common curriculum for all students draws students together; on the other hand, a fragmented curriculum based on tracking separates students by ability or career goals.

In responding to individual differences, the common curriculum leads to unity and builds character among students.

The STEM curriculum is closely coordinated to related subjects such as mathematics and engineering. The mathematics curriculum supports and is closely related to the sciences. The various developmental levels of science are coordinated so what's taught in the third grade science builds on the second grade content and leads to the fourth grade science curriculum.

Instead of offering separate subjects, the more progressive schools include elementary science as part of an integrated curriculum. Selected integration results in better achievement and improved attitudes. The focus of the curriculum on active learning emphasizes results. This means improved learning for all students. A quality learning curriculum is teacher friendly, with clear objectives and less attention given to mindless activities. The new curriculum is also attuned to personal relevance. This includes technology amid the various disciplines, giving students the tools they need to improve society.

Engineering and technology are featured alongside the physical sciences, life sciences, earth and space sciences for two critical reasons: to reflect the importance of understanding the human built world and to recognize the value of better integrating the teaching and learning of science, engineering, and technology.

—National Research Council

THE EXCITING WORLD OF SCIENCE, TECHNOLOGY, AND ENGINEERING EDUCATION

Science, technology, and engineering can be the most exciting experience for elementary and middle school students and teachers if they are taught as active hands-on subjects where students learn through doing. Science, technology, and engineering provide imaginative teachers many opportunities for helping to teach students who struggle with STEM. For one thing, the language used is often difficult to understand.

Too many students still feel that the STEM subjects are either too difficult or too boring. How are teachers going to help students trudge their way through it, or pass the tests? For elementary and middle school students struggling with their science textbooks, those who aren't naturally drawn to the sciences or for whatever reason, and those who can't seem to connect ideas to knowledge comprehension, it is a major problem. We want these students to know they are not alone.

STEM LEARNING STRATEGIES

Many strategies are based on the idea that teachers adapt instruction to student differences. Today, teachers are determined to reach all students, trying to provide the right level of challenge for students who perform below grade level, for gifted students, and for everyone in between. They are working to deliver instruction in ways that meet the needs of auditory, visual, and kinesthetic learners while trying to connect to students' personal interests. Following are some teaching strategies for STEM instruction. They are starting points for consideration, not a complete guide. Feel free to revise and edit the list as you see fit.

1. **Apply a Collaborative Approach**

   Collaborative learning is a "total class" approach that lends itself to STEM instruction. It requires everyone to think, learn, and teach. Within a collaborative learning classroom, there will be many and varied strengths among students. Every student will possess characteristics that will lend themselves to enriching learning for all students. Sometimes, these "differences" may constitute a conventionally defined "disability"; sometimes, it simply means the inability to do a certain life or school-related task; and sometimes, it means, as with the academically talented, being capable of work well beyond the norm. Within the collaborative learning classroom, such exceptionality need not constitute a handicap.

   Collaborative learning is not simply a technique that a teacher can just select and adopt in order to "accommodate" a student within the classroom. Making significant change in the classroom process requires that teachers undergo changes in the ways that they teach, and in the ways they view students. This means creating comfortable, yet challenging, learning environments rich in diversity. The goal is collaboration among all types of learners. In mixed-ability groups, the emphasis must be on proficiency rather than age or grade level, as a basis for student progress.

   Active collaboration requires a depth of planning, a redefinition of planning, testing, and classroom management. Perhaps, most significantly, collaborative learning values individual abilities, talents, skills, and background knowledge.

2. **Form Multiage Flexible Groups**

   To maximize the potential of each learner, educators need to meet each student at his or her starting point and ensure substantial growth during each school term. Classrooms that respond to student differences benefit virtually all students. Being flexible in grouping students gives them many options to develop their particular strengths and encourages them to show their performance.

3. **Set Up Learning Centers**

   A learning center is a space in the class that contains a group of activities or materials designed to teach, reinforce, or extend a particular concept. Centers generally focus on an important topic and use materials and activities addressing a wide range of reading levels, learning profiles, and student interests.

   A teacher may create many centers such as a science center, a music center, or a reading center. Students don't need to move to all of them at once to achieve competence with a topic or a set of skills. Have students rotate among the centers. Learning centers generally include activities that range from simple to complex.

Effective learning centers usually provide clear directions for students including what a student should do if he or she completes a task, or what to do if they need help. A record-keeping system should be included to monitor what students do at the center. An ongoing assessment of student growth in the class should be in place, which can lead to teacher adjustments in center tasks.

4. **Develop Tiered Activities**

These are helpful strategies when teachers want to address students with different learning needs. For example, a student who struggles with reading from a science textbook, or has a difficult time with complex vocabulary, needs some help in trying to make sense of the important ideas in a given chapter. At the same time, a student who is advanced well beyond his or her grade level needs to find a genuine challenge in working with the same concepts.

Teachers use tiered activities so that all students focus on necessary understandings and skills, but at different levels of complexity and abstractness. By keeping the focus of the activities the same, but providing different routes of access, the teacher maximizes the likelihood that each student comes away with important skills and is appropriately challenged (Tomlinson & Moon, 2013).

Teachers should select the concepts and skills that will be the focus of the activity for all learners. The tiered approach includes using assessments to find out what the students need and creating an interesting activity that will cause learners to use an important skill or understand a key idea. It is important to provide varying materials and activities. Teachers match a version of the task to each student based on student needs and task requirements. The goal is to match the task's degree of difficulty and the student's readiness.

5. **Make learning more challenging**

Challenging strategies put more emphasis on authentic problems where students are encouraged to formulate their own problem on a topic they're interested in, and work together with other students to solve it. Problems chosen by students should be connected to the "real-world" and should allow time for discussion and sharing of ideas among students.

6. **Have a clear set of standards**

Integrating standards into the curriculum helps make learning more meaningful and interesting to reluctant learners. Having a clearly defined set of standards helps teachers concentrate on instruction and makes the expectations of the class clearer to the students. Students come to understand what is expected and work collaboratively to achieve it. Challenging collaborative groups to help each other succeed is another way to avoid poor performance.

7. **Expand learning options**

Not all students learn in the same way or at the same time. Teachers can expand learning options by differentiating instruction. This means teachers should be reaching out to struggling students or small groups to improve teaching in order to create the best learning experience possible.

8. **Introduce Active Reading Strategies**

There is an approach which uses "active reading" strategies to improve students' abilities to explain difficult text. This step-by-step process involves reading aloud to yourself or someone else, as a way to build science understandings. Although most learners self-explain without verbalizing, the active reading approach is similar to that used by anyone attempting to master new material: the best way to truly learn is to teach and explain something to someone else.

THE CHANGING SCIENCE, TECHNOLOGY, AND ENGINEERING CURRICULUM

Today, active science learning in the elementary and middle schools is changing the boring textbook process. It contributes to the development of interdisciplinary skills. For example, the overlap among science, technology, engineering, and mathematics is obvious when you look at common skills. Many of the best models in science education involve having students work in cross-subject and mixed-ability teams. Teachers begin by making connections between the STEM subjects and the rest of the curriculum. Introducing real-world concerns helps; good examples might be found in the newspaper.

One thing is for sure: to use and understand the STEM subjects today requires an awareness of what's going on and how it relates to our culture and our lives.

Technology, engineering, and mathematics are tools of science. It would be a mistake to simply attach these dynamic tools to the old systems of instruction like those involving lecture content, basic drills, homework, and over contrived problems.

Put the latest tools in the hands of skilled teachers and motivated students and good things can happen. A lot can be done to encourage inquiry and problem solving by generating curiosity and observation, posing questions, and actively seeking answers. Technology, for example, can turn up the volume, but it all comes down to clear goals, training good teachers, and the quality of the overall classroom environment.

THE NATIONAL SCIENCE AND STEM CONTENT STANDARDS

The content standards clearly articulate what students should know, understand, and be able to do. Students should be able to:

- Understand the basic concepts and processes in science and STEM instruction.
- Use the process of inquiry when doing science, technology, engineering, and math.
- Apply the properties of physical science, life science, earth and space science when doing activity-based learning.
- Use STEM understandings to design solutions to problems.
- Understand the connections between science and technology.
- Examine and practice science and STEM from personal and social viewpoints.
- Identify with the history and nature of science, technology, and mathematics through readings, discussions, observations, and writings (National Research Council, 1996).

INQUIRY IN THE SCIENCE STANDARDS AND THE PROCESS SKILLS

The inquiry skills of science and STEM are acquired through a questioning process. Inquiry also raises new questions and directions for examination. The findings may generate ideas or suggest connections or ways of expressing concepts and interrelationships more clearly. The process of inquiry helps students grow in content knowledge and the processes and skills of the search. It also invites unmotivated learners to explore anything that interests them.

Whatever the problem, subject, or issue, any inquiry that is done with enthusiasm and with care will use some of the same thinking processes that are used by scholars who are searching for new knowledge in their field of study.

Inquiry processes form the foundation for understanding and are components of the basic goals and standards of science and mathematics. These goals are intertwined and multidisciplinary, providing students many opportunities to become involved in inquiry. Each goal involves one or more processes (or investigations).

The inquiry process approach includes the major process skills and standards as outlined in the activities that follow. The science activities also include the key principles of a STEM classroom. These include **content**: what students will learn; **process**: the activities by which students make sense of important ideas using necessary skills; **product**: how students show what they have learned and prove their point; and **learning environment**: safe comfortable conditions that set the tone for learning (Tomlinson, 2014; Tomlinson & Cunningham Edison, 2003).

SCIENCE, TECHNOLOGY, AND ENGINEERING ACTIVITIES BASED ON THE SCIENCE STANDARDS

This section connects the science standards to elementary and middle school classrooms. The importance of establishing activities that use the inquiry skills of observing, measuring, recording data, and drawing reasonable conclusions is emphasized. Whenever possible, mathematics is included in the activities so that math and science skills are developed together. Careful attention has been given to the sequence of activities within each section; the more general introductory activities come first, followed by more focused activities that build on each other to develop student understanding. At the end of each activity, suggestions for STEM instruction are offered. These ideas provide "a peek" into the STEM process so that teachers can try out some differentiated strategies with their students.

SCIENCE, TECHNOLOGY, AND ENGINEERING ACTIVITIES

Activity 1: Buttons and Shells (grades K–5)

Inquiry Skills: observing, classifying, comparing, sequencing, solving problems, group work, communicating, recording, gathering data, measuring.

Science and STEM Standards: inquiry, physical science, science and technology, personal perspectives, written communications.

Engage: In this introductory activity, students are observing by looking at the details of an object very closely. Students collect evidence by classifying, comparing, sequencing, gathering data, and trying to solve problems. Students communicate and work in groups to arrive at a solution.

New Vocabulary: sorting, striation, gradation, Venn diagrams.

Materials: bags of assorted buttons (one for each group); yarn or string dividers; bags of small sea shells (one for each group).

Problem 1: How many ways can you sort your bag of buttons? Try to sort them at least ten different ways.

Problem 2: Make a Venn diagram using your bag of buttons.

Venn Diagram: A method of illustrating set unions and intersections. For example, a set of blue buttons is one category; a set of round buttons is another category. A set of blue round buttons is an intersecting category.

Problem 3: Classify the shells by striation: light to dark (color), small to large, number of ridges, amount of water shell can hold.

Ways to Differentiate Instruction

The students work in groups to sort and classify the buttons ten different ways. The teacher collects learning materials making sure students can classify, sort, and find as many ways as possible. The teacher adapts instruction when students are to create a Venn diagram. The teacher wants students to have as many chances as possible to find patterns, make comparisons, and figure out what intersection means. The teacher models examples that help students distinguish what is similar and different and encourages partnerships that build success. In the final problem-serrating shells, similar strategies are employed.

Activity 2: Unknown Liquids: Experiment (grades 3–6)

Inquiry Skills: hypothesizing, experimenting, and communicating.

Science and STEM Standards: inquiry, physical science, science and technology, personal perspectives, written communications.

Description: In this exploring activity, students are experimenting with chemicals and doing physical science work. They are learning to use tools found in the lab and becoming familiar with the safety rules of science, mathematics, and technology.

Materials for each table:

5 plastic containers
5 liquids (oil, water, soap, alcohol, vinegar)
5 medicine droppers
1 beaker
1 small plastic beaker
1 tray
1 sheet of plastic wrap
1 sheet of aluminum foil
1 sheet of waxed paper

Engage:

1. Set up the containers with liquids.
2. Discuss the colored liquids.
3. Set up the trays with papers and materials for each group: five eye droppers, two beakers, one small plastic container.
4. On the board, list some possible experiments with liquids:
   • liquid races
   • floatability
   • density
   • mixing liquids
   • other ideas

Explore: Try to discover what the four liquids are.

Rules:

1. Each liquid is a household substance that may or may not have been colored with food coloring to hide its identity.
2. Students are limited to using only their sense of sight to do this experiment. Students should experiment by manipulating the liquids.
3. For safety reasons, caution students they are not to smell, touch, or taste the chemicals.
4. Each medicine dropper may be used to pick up only one liquid. We do not want contamination!

Explain: Try to find out what the four unknown liquids are.

1. Write a description of the activity.
2. Explain how the group went about solving the problem.

Elaborate:

1. Write what the group learned from the activity.

Evaluate:

1. Provide some follow-up suggestions of how the activity could be improved or give suggestions of what to do after the activity was finished.

Suggestions for STEM Instruction

The students work together to try and figure out what the unknown liquids are. The teacher encourages a collaborative approach. He/she offers suggestions of ways students can apply the knowledge of what they know about common chemicals. He/she suggests:

• testing different variables such as dropping a chemical on each sheet of paper to see how the papers react to the chemicals,

- shake the containers with the liquids,
- try some possible experiments,
- observe, test, and be creative in your approach.

Activity 3: Demonstrating the Behavior of Molecules (grades 3–6)

Inquiry Skills: observing, comparing, hypothesizing, experimenting, and communicating.

Science and STEM Standards: inquiry, physical science, science, technology, personal perspectives, written communications.

Background Information Description: This activity simulates how molecules are connected to each other and the effects of temperature change on molecules.

Engage: Students usually have questions about the way things work. The questions students naturally ask are, for example, "why does ice cream melt?" "why does the tea kettle burn my hand?" "where does steam come from?" and "why is it so difficult to break rocks?"

Explain: Explain that molecules and atoms are the building blocks of matter. Heat and cold energy can change molecular form.

Explore: The class is, then, asked to participate in the "hands-on" demonstration of how molecules work. This is a great opportunity for students to participate and perhaps, assume leadership as a group leader. Before beginning the demonstration, explain that matter and energy exist and can be changed, but not created or destroyed.

Elaborate: Ask for volunteers to role play the parts of molecules. Direct students to join hands showing how molecules are connected to each other, explaining that this represents matter in a solid form. Next, ask them to "show what happens when a solid becomes a liquid." Heat causes the molecules to move more rapidly so they can no longer hold each other together. Students should drop hands and start to wiggle and move around. The next question, "how do you think molecules act when they become a gas?"

Evaluate: Carefully move students to the generalization that heat transforms solids into liquids and then, into gases. The class enjoys watching the other students wiggle and fly around as they assumed the role of molecules turning into a gas. The last part of the demonstration was the idea that when an object is frozen, the molecules have stopped moving altogether. The demonstration and follow-up questions usually spark a lot of discussion and more questions.

Suggested Ways to Individualize Instruction

Interest and student motivation are paramount in this activity. Challenge students to discover how molecules are everywhere. The students are part of a hands-on demonstration trying to answer their questions about molecules. The teacher modifies instruction when she asks for volunteers to play the role of molecules even though she wants all students to participate. Collaborative group work, discussions, and lots of communication take place in this STEM hands-on activity.

Activity 4: What Will Float? (grades 3–6)

Inquiry Skills: hypothesizing, experimenting, and communicating.

Science and STEM Standards: inquiry, physical science, science and technology, personal perspectives, written communications.

Description: The weight of water gives it pressure. The deeper the water, the more pressure. Pressure is also involved when something floats. For an object to float, opposing balanced forces work against each other. Gravity pulls down on the object, and the water pushes it up. The solution to floating is the object's size relevant to its weight. If it has a high volume and is light for its size, then, it has a large surface area for the water to push against. In this activity, students will explore what objects will float in water. All students should try to float some of these objects.

Materials: large plastic bowl or aquarium, salt, bag of small objects (paper clip, nail, block, key, etc.) to test, ruler, spoon, oil base modeling clay, paper towels, large washers, kitchen foil 6-inch square.

Engage:

1. Have the students fill the plastic bowl half full with water.
2. Direct the students to empty the bag of objects onto the table along with the other items.

Explore:

1. Next, have students separate the objects into two groups: the objects that will float and the objects that will sink.

Explain:
1. Encourage students to record their predictions in their STEM journal.

Elaborate:

1. Have students experiment by trying to float all the objects and record what happened in their STEM journals.

Evaluation: Have students reflect on these thinking questions and respond in their STEM journals. Encourage students to work together, helping students who are having trouble expressing their ideas.

1. What is alike about all the objects that floated? Sank?
2. What can be done to sink the objects that floated?
3. What can be done to float the objects that sank?
4. In what ways can a piece of foil be made to float? Sink?
5. Describe how a foil boat can be made.
6. How many washers will the foil boat carry?
7. What could float in salt water that cannot float in fresh water?
8. Encourage students to try to find something that will float in fresh water and sink in salt water.

STEM Instruction Suggestions

Spell out the purpose of the activity. Students have fun experimenting with what will float. Students construct boats made from aluminum foil. This motivating activity looks at water pressure, gravity, volume, weight, and ways to solve problems. The teacher individualizes by making it clear what students are to learn; she understands, appreciates, and builds on student differences, adjusting content, process, and product in response to student readiness, interests, and learning profile.

Activity 5: Water Cohesion and Surface Tension (grades 2–5)

Inquiry: hypothesizing, experimenting, and communicating.

Science, Technology, and Engineering Standards: earth science, science, engineering and technology, personal perspectives, written communications.

Description: Students will determine how many drops of water will fit on a penny in an experiment that demonstrates water cohesion and surface tension.

Materials:

- one penny for each pair of students
- glasses of water
- paper towels
- eye droppers (one for each pair of students)

Engage: Have students work with a partner. As a class, have them guess how many drops of water will fit on the penny.

Explore: As a class, have them guess how many drops of water will fit on the penny.

Explain: Record their guesses on the chalkboard.

Elaborate: Ask students if it would make a difference if the penny was heads or tails. Also, record these guesses on the chalkboard. Instruct the students to try the experiment by using an eyedropper, a penny, and a glass of water. Encourage students to record their findings in their STEM journal. Bring the class together again. Encourage students to share their findings with the class. Introduce the concept of cohesion. (Cohesion is the attraction of like molecules for each other. In solids and liquids, the force is strongest. It is cohesion that holds a solid or liquid together. There is also an attraction among water molecules for each other.) Introduce and discuss the idea of surface tension. (The molecules of water on the surface hold together so well that they often keep heavier objects from breaking through. The surface acts as if it is covered with skin.)

Evaluation, Completion, and /or Follow-up: Have students explain how this activity showed surface tension. Instruct students to draw what surface tension looked like in their science journal. What makes the water drop break on the surface of the penny? (It is gravity.) What other examples can students think of where water cohesion can be observed? (Rain on a car windshield or window in a classroom, for example.) Even disinterested students can relate to this activity if drawn into the conversation.

STEM Instruction Suggestions

The teacher individualizes by modifying instruction based on her ongoing assessment of students' science, technology, and engineering knowledge. He/she explains what students are to learn and gives them opportunities to work with a partner. Students observe, ask questions, discuss, and record their findings. Writing about surface tension is an important follow-up activity.

Activity 6: Experimenting with Surface Tension (grades 3–8)

Inquiry Skills: hypothesizing, experimenting, and communicating.

Science Standards: inquiry, physical science, science and technology, personal perspectives, and written communications.

Engage: Students will develop an understanding that technological solutions to problems, such as phosphate-containing detergents, have intended benefits, and may have unintended consequences.

Explore: Students apply their knowledge of surface tension. This experiment shows how water acts like it has a stretch skin because water molecules are strongly attracted to each other. Students will also be able to watch how soap molecules squeeze between the water molecules, pushing them apart and reducing the water's surface tension.

Background information: Milk, which is mostly water, has surface tension. When the surface of milk is touched with a drop of soap, the surface tension of the milk is reduced at that spot. Since the surface tension of the milk at the soapy spot is much weaker than it is in the rest of the milk, the water molecules elsewhere in the bowl pull water molecules away from the soapy spot. The movement of the food coloring reveals these currents in the milk.

Engage: Divide class into groups of four or five students.

Materials: Milk (only whole or 2 percent fat milk will work), newspapers, a shallow container, food coloring, dish washing soap, a saucer or a plastic lid, toothpicks.

Elaborate:

1. Take the milk out of the refrigerator 1/2 hour before the experiment starts.
2. Place the dish on the newspaper and pour about 1/2 inch of milk into the dish.
3. Let the milk sit for a minute or two.

Explore:

1. Near the side of the dish, put one drop of food coloring in the milk. Place a few colored drops in a pattern around the dish. What happened?
2. Pour some dish washing soap into the plastic lid. Dip the end of the toothpick into the soap, and touch it to the center of the milk. What happened?

Explain:

1. Dip the toothpick into the soap again, and touch it to a blob of color. What happened?
2. Rub soap over the bottom half of a food coloring bottle. Stand the bottle in the middle of the dish. What happened?

Elaborate:

1. The colors can move for about twenty minutes when students keep dipping the toothpick into the soap and touching the colored drops.

Evaluation: Students will discuss their findings and share their outcomes with other groups. Using stations adds another dimension. Have students record their findings and explain why their results turned out as they did.

Suggestions for STEM Instruction

The teacher attends to students' interests, learning styles, prior needs, and comfort zones. In this activity, students are to create and test ramps using a variety of objects. Each group has a chance to predict which slide will reach the barrier. Students will record their test trials and reflect on their results. A teacher can modify the problem and differentiate according to individual needs. If a student couldn't get her ramp to work, the teacher should ask the class to assist her in fixing it. Students should share their reflections and successes.

Activity 7: Recyclable Materials Construction (grades 6–9)

Science standards: science inquiry, physical science, science and technology, math and science, coordination, and problem solving.

Inquiry Skills: observation, prediction, measurement, data recording.

Description: "Hands-on technology" is the exciting things that happen during technological problem solving when students develop and construct their own "best" solution. This middle school activity moves beyond conducting experiments or finding solutions to word problems (all students doing the same task at the same time). With "hands-on technology," students are not shown a solution. Typically, this results in some very creative designs.

Engage: Using the tools and materials found in a normal middle school technology education laboratory, students will design and construct solutions that allow them to apply the process skills. The products they create and engineer in the technology lab often use a wide range of materials such as plastics, woods, electrical supplies, and so on. During the course of solving their problem, students will be forced to test hypotheses and frequently generate new questions. This involves a lot of scientific investigation and mathematical problem solving, but it is quite different from the routine classroom tasks. In this activity, a problem is introduced to the class.

Explore: Working in small groups of four or five students, pose a challenge to students to plan a way of coming up with a solution. Students are to document the steps they used along the way. Some suggestions: have students brainstorm and discuss with friends, draw pictures, show design ideas, use mathematics, present technical drawings, work together, and consult with experts.

Elaborate: The best construction materials are strong, yet lightweight. Wood is unexpectedly strong for its weight, and therefore, well suited for many structures. Larger buildings often use steel reinforced concrete beams, rather than wood, in their construction. However, steel and concrete are both heavy, presenting problems in construction. A lighter material would be a great alternative and a best seller in the construction industry. This could be done by reinforcing the beam with a material other than steel—ideally, a recyclable material.

Evaluate: Design the lightest and strongest beam possible by reinforcing concrete with one or more recyclable materials: aluminum cans, plastic milk jugs, plastic soda bottles, and/or newspaper. Students must follow the construction constraints. The beam will be weighed. Then, it will be tested by supporting it at each end, and a load will be applied to the middle. The load will be increased until the beam breaks. The load divided by the beam weight will give the load-to-weight ratio. The designer of the beam with the highest load-to-weight ratio will be awarded the contract.

Construction Limits:

The solution must:

1. Be made into a reusable mold that the students design.
2. The result should be a 40 centimeter (approximately 16 inches) long beam that fits within a volume of 1050 cubic centimeters (approximately 64 cubic inches).
3. Be made from concrete and recyclable materials.

Engage:

1. Groups of students will plan and design their beam.
2. Groups will work on their construction plans.
3. Students will design and construct their beam.
4. Students will gather information from a variety of resources and make sketches of all the possibilities they considered.
5. Students will record the science, mathematics, and technology principles used.

Explore:

1. Divide students into small groups of three or four students.
2. Present the problem to the class.
3. Students will discuss and draw out plans for how to construct a beam. All students should be part of this process.
4. Students will design a concrete beam reinforced with recycled materials.
5. Students will work together to construct, measure, and test the beam.

Explain:

1. Students will present their invention to the class.

Elaborate/Evaluate: Students will document their work in a portfolio that includes:

1. Sketches of all the possibilities their group considered.
2. A graphic showing how their invention performed.
3. Descriptions of the process skills used in their solution.
4. Information and notes gathered from resources.
5. Thoughts and reflections about this project. Some students may need assistance in their designs. Encourage them to work together on their construction.

STEM Suggestions

Teachers can use tiered activities so all students focus on basic understandings and skills but at different levels of complexity. By keeping the focus of the activity the same, but offering ways of access at varying degrees of difficulty, the teacher gets the most out of each student so that each student comes away with essential skills and each student is appropriately challenged. Another way to differentiate is to modify students' products (in this case, constructions). Compare students who have structures that are more concrete to those whose structures are quite complex.

Suggestions for Teachers

Nearly all educators agree with the goal of STEM instruction, but teachers may not have strategies for making it happen. Here are a few hints that teachers can use to enhance instruction:

1. **Assess students.**
   The role of assessment is to foster worthwhile learning for all students. Performance assessments, informal assessment tools such as rubrics, checklists, and anecdotal records are some assessment strategies that are helpful for students with learning problems. Teachers may use a compacting strategy. This strategy encourages teachers to assess students before beginning a unit of study or development of a skill.
2. **Create complex instruction tasks.**
   Complex tasks are:
   • Open-ended,
   • intrinsically interesting to students,
   • uncertain (thus allowing for a variety of solutions),

- involve real objects,
- draw upon multiple intelligences in a real-world way.

3. **Use television in the classroom.**

Television's wide accessibility has the potential for making learning available for students who do not perform well in traditional classroom situations. It can reach students on their home ground, but the most promising place is in the classroom.

4. **Use materials and activities that address a wide range of reading levels, learning profiles, and student interests.**

Include activities that range from simple to complex, from concrete to abstract.

5. **Use STEM notebooks.**

Science notebooks are an everyday part of learning. The science notebook is more than a record of collected data and facts of what students have learned. They are notebooks of students' questions, predictions, claims linked to evidence, conclusions, and reflections.

A STEM notebook is a central place where language, data, and experiences work together to produce meaning for the students. Notebooks support STEM learning. They are helpful when addressing the needs of disinterested students. In a science notebook, even students who may have poor writing skills can use visuals such as drawings, graphs, and charts to indicate their learning preferences. There is ongoing interaction in the notebooks. For teachers, a notebook provides a window into students' thinking and offers support for all students.

6. **Provide clear directions for students.**

Teachers need to offer instructions about what a student could do if he or she needs help.

7. **Use a record-keeping system to monitor what students do.**

8. **Include a plan for ongoing assessment.**

Teachers use ongoing assessment of student readiness, interest, and learning profile for the purpose of matching tasks to students' needs. Some students struggle with many things, others are more advanced, but most have areas of strengths. Teachers do not assume that one set of skills fits all students.

9. **Modify curricular elements.**

Adjust content, process, and products only when you see a student in need and you are convinced that the learner will understand important ideas and skills more thoroughly as a result.

As an extension of the STEM subjects, teachers must be prepared to engage their students in frequent reading, writing, speaking, and listening. The full range of language tools is needed if students are going to reflect, inquire, problem solve, and communicate what they are learning. The basic idea is to empower children in ways that help them construct their own knowledge.

To paraphrase Nietzsche:
Meaning resides in the rainbow colors
around the outer edges of the imagination and knowledge.

PREPARING INFORMED AND CREATIVE CITIZENS

In an increasingly scientific and technological world, everyone needs some basic competencies in the STEM subjects. As a consequence, an important educational goal is to prepare citizens who are aware of the implications of science, technology, engineering, and mathematics. This means preparing students who can make use of knowledge and connect the implications of science, technology, engineering, and math in their personal lives and in society.

STEM literacy involves having a broad familiarity with related issues and the key concepts that underlie them. At school, this means organizing inquiry and problem solving around real-life problems—the kind that can elicit critical thinking and shared decision-making. The process also involves curiosity, observation, posing questions, and actively seeking answers.

The recognized importance of a STEM literate citizenry has resulted in national efforts to reform instruction in these subjects. Strategies include concrete, physical experiences and opportunities for students to explore science, technology, engineering, and mathematics in their lives. There is an increased emphasis on ideas and thinking skills. This involves sequencing instruction from the concrete to the abstract. Also, attention must be given to developing effective oral and written communication skills.

Frequent group activity sessions are provided where students are given many opportunities to question data, to design and conduct real experiments, and carry their thinking beyond the class experience. As students raise questions that are appealing and familiar to them, activities arise, which improve reasoning and decision-making. Collaborative learning has become the primary grouping strategy where learning is done as a cohesive group in which ideas and strengths are shared.

STEM can be an exciting experience for students and teachers when it is taught as an active hands-on subject. Connecting with other disciplines can provide many opportunities for integration with other subjects. Teachers need subject matter knowledge that is broad and deep enough to work with second language learners and others who may have difficulty with their school work.

This often requires improving language and broad-based literacy development possibilities to get at content.

By focusing on real investigations and participatory learning, teachers can move students from the concrete to the abstract as they explore themes that connect science, technology, engineering, and mathematics. As advances are made in these subjects, a whole series of moral dilemmas are generated (Wilson, 2014). As we move along a path to the future, it would be naive to assume that material advances will make us better people.

Teaching strategies include many participatory experiences and opportunities for students to explore how the STEM approach could play a role in their lives. The emphasis on inquiry and problem solving involves posing questions, conducting investigations, providing explanations, and communicating the results. Setting up moral dilemmas for student debate can make for lively discussions and deeper understandings.

Students develop effective interpersonal skills as they work together, pose questions, and critically examining data. This often means designing and conducting real experiments that carry their thinking beyond the classroom. As instruction becomes more connected to students' lives, enriching possibilities arise from inquiring about real-world concerns.

All students can learn science, technology, engineering, and mathematics, and they should have the chance to become scientifically literate. For example, one of the themes in the National Science Education Standards emphasizes the processes of science. Also, the science standards give a great deal of attention to cognitive abilities such as logic, evidence, and extending their knowledge to construct explanations of natural phenomena.

Young students are naturally curious and eager to explore. So, STEM literacy should begin in the early grades. This means encouraging students to get actively involved in actually *doing things* that move them along the road to awareness.

SUMMARY, CONCLUSION, AND LOOKING AHEAD

The quality of STEM instruction has a lot to do with how we teach and why we teach it. Related research and the content standards suggest that it is important to look at how, why, and where the subjects of science, technology, engineering, and mathematics work. Remember, changes in school practice involve educators communicating good ideas and encouraging a social environment.

For students, it is important to build up an information base while developing powers of mind. The objective is to develop and deepen the understanding of science, technology, engineering, and mathematics. Ideas and concepts gleaned from these subjects matter because they are a major source of decisions and actions in the modern world (Montgomery & Chirot, 2015).

Teachers who prepare lessons in science, technology, engineering, and math can make those subjects more meaningful by involving students in activities that relate to real-life situations. Active learning can help learners make connections with the past, the present, and the rest of the world.

Along the path to literacy in the STEM subjects, it is possible to reflect human values and emphasize both personal and social responsibility. In today's collaborative classrooms, unity and diversity can go together. Self-understanding and responsible citizenship involve using knowledge to make wise decisions and solve interrelated problems related to life and living.

Learning at all levels is becoming more interdisciplinary. For example, at the highest levels of science, new interrelated and emerging research fields topics include biochemistry, biophysics, plant engineering, terrestrial biology, and neurobiology, to name a few. Even at the primary school level, the STEM subjects have dimensions that extend to the arts, the social sciences, and topics ranging from ethics and values to law and social justice.

To maximize learning in a collaborative classroom, it helps when teachers can sometimes work together to develop lessons that build on relationships between the concepts covered and life experiences. Connected autonomy, feedback, and diffusion are all a part of the process. As teachers are given more flexibility, it is important that they learn to strike a balance between autonomy and collaboration.

Schools where teachers cooperate with each other have a better chance of making the adaptive changes needed to meet student needs (Fullan, 2015).

Many educators recognize that the time has come to develop curricula in a way doesn't isolate the STEM subjects from other subjects and the socioeconomic realities of today's world. Of course, it takes more than good intentions to make things happen. It all boils down to improving instruction and the home environment in ways that help all children achieve their full potential (Russakoff, 2015).

A vital function of schools is to help students develop as creative and critical thinkers who can analyze elements of the society in which they live. In the classroom, this means intellectually engaging with today's problems and ideas. Lessons must contribute to student's developing habits of the mind. This must all be done in a context of natural realities, technological forces, and the opportunities that may be confronted in the future.

When thinking about what's out there over the horizon, it is important to remember that there are fragments of the future sprinkled around today.

History may or may not repeat itself, but it often rhymes with the past. So, there are times when looking at how issues came up and were handled yesterday can be helpful when it comes to dealing with the problems of today and tomorrow.

REFERENCES

Brandt, R.S. (Ed.) (2000). *Education in a new era: 2000 ASCD yearbook.* Alexandria, VA: Association for Supervision and Curriculum Development.

Firestein, S. (2012). *Ignorance: How it drives science.* Oxford, UK: Oxford University Press.

Fullan, M. (2015). *Freedom to change.* San Francisco, CA: Jossey-Bass.

Montgomery, S. & Chirot, D. (2015). *The shape of the new: Four big ideas and how they made the modern world.* Princeton, NJ: Princeton University Press.

National Research Council (2012). *A framework for K–12 science education: Practices crosscutting concepts, and core ideas.* Alexandria, VA: Association for Supervision and Curriculum Development (National Assessment Governing Board, 2011,xi, 1–4; National Research Council 2012).

Russakoff, D. (2015). *The prize: Who's in charge of America's schools.* New York, NY: Houghton Mifflin Harcourt.

Tomlinson, C. (2014). *The differentiated classroom: Responding to the needs of all learners* (2nd edition). Alexandria, VA: Association for Supervision and Curriculum Development.

Tomlinson, C. & Cunningham Edison, C. (2003). *Differentiation in practice: A resource guide for differentiating curriculum.* Alexandria, VA: Association for Supervision and Curriculum Development.

Tomlinson.,C. & Moon, T. (2013). Assessment and student success in a differentiated classroom. Alexandria, VA: Association for Supervision and Curriculum Development.

Vasquez, J, Sneider, C., & Comer, M. [Foreword by R. Bybee] (2013). *STEM lesson essentials, grades 3–8: Integrating science, technology, engineering, and mathematics.* Portsmouth, NH: Heinemann.

Wilson, E. (2014). *The meaning of human existence.* New York, NY: Liveright Publishing (W.W. Norton).

Chapter Three

Science and Mathematics

The Power of Inquiry and Problem Solving

In many ways STEM is not a new subject because the fields involved have been part of the curriculum for decades. What's new is asking students to make additional connections and apply some of the concepts learned across the curriculum.

Many teachers of science and mathematics at the elementary school level have to teach dozens of topics and subjects. So, it should not come as a surprise to find out that some have not received extensive training in every subject.

When it comes to STEM teaching, it is important for all teachers to have the intellectual tools needed to help children learn and apply age-appropriate concepts. Teachers can make a difference. But policy prescriptions have to go beyond curriculum and instruction to deal with social forces that reproduce inequality across generations.

Learning about the STEM subjects isn't easy; sometimes, it is like exploring an unknown city where you can ever be sure what is just around the corner. The world is an ambiguous place. But it is one thing to realize the limits of predictability and quite another to dismiss all prediction as a waste of time (Tetlock & Gardner, 2015).

Propelling students' academic life forward requires preparing them for chance encounters with the unknown. Major goals of science and math education now include helping students prepare for future discoveries and changes. This also means helping students develop more social skills and self-understanding as they move in the direction of responsible citizenship.

The problems of today—and in the future—require not only STEM experts, but also citizens who can understand the implications of these subjects. By working together on real-world problems, students can advance their understanding of subject matter and creative thinking. And remember, many new approaches and innovative ideas are the result of intellectual curiosity.

Science, technology, engineering, and mathematics are so well connected that it can sometimes be hard to figure out where one field starts and another ends. Advances in any combination of these fields generate a whole series of moral dilemmas. Along the glide path to deeper learning, it also is important that learners recognize how new ideas generated by these subjects are altering the way knowledge is conveyed (Barnes, 2013).

CORE CURRICULUM STANDARDS FOR SCIENCE AND MATHEMATICS

The National Research Council (NRC) has published *A Framework for K–12 Science Education: Practices Crosscutting Concepts and Core Ideas* (2011). This publication takes one of the most important steps forward in science and math education since the *National Science Standards* and *The National Mathematics Standards*. The NRC constructed the science framework and is working on the next generation of science standards.

The standards bring an imaginative perspective to learning science and math. They emphasize the importance of challenging students in *doing* science and mathematics, not just learning content.

The science framework identifies key scientific ideas and practices that all students should learn. Among other things, it is designed to help students gradually develop their knowledge of core ideas in four interdisciplinary areas over multiple years, rather than the shallow knowledge of focusing on many topics.

The latest science framework strongly emphasizes the ways students carry out science investigations and encourages arguments based on evidence (National Academy Press, 2011).

The math standards merge the common core standards grades K–12. New to the 2011 math framework are the standards for practical practice. They describe math-proficient students, presenting a coherent progression and a strong foundation that prepares students for future math classes.

The goal of the new standards is to ensure that all students have an appreciation of the beauty and wonder of science, technology, and mathematics. Students should have the capacity to discuss and think critically about science-, engineering-, and math-related issues and pursue careers in science, math, and engineering. Currently, science and math in the United States lack a common vision of what students should know and be able to do (Bybee, 2013).

Over the last few decades, both teaching methods and subject matter content have changed. So, have the textbooks. Increasingly, students have e-textbooks, e-books, e-zines that can be accessed on all kinds of mobile devices. Still, hard-copy books have a role to play in any pedagogical version of the future. It is up to educators to decide the technological mix that works best in different situations.

STEM ACTIVITIES AND THE SCHOOL CURRICULUM

The evolving nature of science, technology, engineering, and mathematics is one of the reasons for so much debate within the scientific community and in the general public. As far as classroom-based discussion, experiments, and computations are concerned, students must learn how to collaboratively put the skills they learn into practice. All of these points connect to the subject matter standards.

Science education as an example—the science interdisciplinary areas include: life science, physical science, earth/space science, engineering, technology, and the application of science. Some core ideas that cut across these fields include, matter and interactions, and energy. Students understand the same is relevant in many fields. These concepts should become familiar as students progress from Kindergarten through grade twelfth.

Both the science and the math standards emphasize key practices that students should learn: asking questions, defining problems, and analyzing and interpreting data. Other important practices include explaining ideas and designing solutions. These practices need to be linked with the study of interdisciplinary core ideas and applied throughout students' education.

The math framework merges the common core standards for grades K to 12. New to the 2011 math framework are the standards for practice. They describe math proficient students, presenting a coherent progression and a strong foundation that prepare students for the 2011 Algebra 1 course (Massachusetts Department of Elementary and Secondary Education, 2010). It emphasizes grade level content standards pre-K to 8. There is a greater focus on improving math achievement.

Guiding Principles for Mathematics

1. *Learning:* Math ideas should be explored in ways that stimulate curiosity, create enjoyment of math, and develop a depth of understanding. Students should be actively engaged in doing meaningful mathematics, discussing ideas and applying math in interesting, thought-provoking situations.

2. Math tasks should be designed to challenge students—for example, some short- and long-term investigations that connect procedures and skills with conceptual understandings. Tasks should generate active classroom talk, promote conjectures, and lead to the understanding of the necessity of math reasoning.

Standards for Math Practice

1. Make sense of problems and be persistent in trying to solve them.
2. Reason, draw conclusions.
3. Create reasonable arguments.
4. Model with math and science situations.
5. Use appropriate tools.
6. Be precise.
7. Clearly express your reasoning.
8. Interpret results.
9. Report on the conclusions and the reasoning behind them.

MATHEMATICS: A TOOL OF SCIENCE

Children have a natural curiosity when it comes to using science and math to examine the natural world. They learn by experiencing things for themselves, building on what they have already learned, and talking with other students about what they are doing. Observing, classifying, measuring, and collecting are just a few examples of the processes that children of all ages can learn and apply (Leinwand, 2012).

Understanding Science and Mathematics

1. Science and math are methods of thinking and asking questions.
   How students make plans, organize their thoughts, analyze data, and solve problems is *doing* science and mathematics. People comfortable with science and math are often comfortable with thinking. *The question* is the cornerstone of all investigation. It guides the learner to a variety of sources revealing previously undetected patterns. These undiscovered openings can become sources of new questions that can deepen and enhance learning and inquiry. Questions such as "How can birds fly?" "Why is the sky blue?" or "How many?" have been asked by children throughout history. Obviously, some of their answers were wrong. But the important thing is that the children never stopped asking—they saw, wondered, and sought an answer.

2. **Science and math both require knowledge of patterns and relationships.**
   Children need to recognize the repetition of science and math concepts and make connections with ideas they know. These relationships help unify the science and math curriculum as each new concept is interwoven with former ideas. Students quickly see how a new concept is similar or different from others already learned. For example, young students soon learn how the basic facts of addition and subtraction are interrelated ($4 + 2 = 6$ and $6 - 2 = 4$). They use their science observation skills to describe, classify, compare, measure, and solve problems.

3. **Science and math are tools.**
   Mathematics is the tool scientists and mathematicians use in their work. It is also used by all of us every day. Students come to understand why they are learning the basic science and math principles and ideas that the school curriculum involves. Like mathematicians and scientists, they also will use science and mathematics tools to solve problems. They will learn that many careers and occupations are involved with the tools of science and mathematics.

4. **Science and math are fun (a puzzle).**
   Anyone that has ever worked on a puzzle or stimulating problem knows what we're talking about when we say science and mathematics are fun. The stimulating quest for an answer prods one on toward finding a solution.

5. **Science and math are art forms.**
   Defined by harmony and internal order, science and mathematics need to be appreciated as art forms where everything is related and interconnected. Art is often thought to be subjective, and by contrast, objective science and mathematics are often associated with memorized facts and skills. Yet, the two are closely related to each other. Students need to be taught how to appreciate the scientific and mathematical beauty all around them—for example, exploring fractal instances of science and math in nature. (A fractal is a wispy tangled curve that seems complicated no matter how closely one examines it. The object contains more, but similar, complexity the closer one looks.) A head of broccoli is one example. If you tear off a tiny piece of the broccoli and look at how it is similar to the larger head you will soon notice that they are the same. Each piece of broccoli could be considered an individual fractal or a whole. The piece of broccoli fits the definition of fractal appearing complicated; one can see consistent repetitive artistic patterns.

6. **Science and math are communication languages.**
   Science and mathematics require being able to use special terms and symbols to represent information. This unique language enhances our ability to communicate across the disciplines of technology, statistics, and other

subjects. For example, a young child encountering $3 + 2 = 5$ needs to have the language translated into terms he or she can understand. Language is a window into students' thinking and understanding.

Our job as teachers is to make sure students have carefully defined terms and meaningful symbols. Statisticians may use mathematical symbols that seem foreign to some of us, but after taking a statistics class, we, too, can decipher the mathematical language. It's no different for children. Symbolism, along with visual aids such as charts and graphs, is an effective way of expressing science and math ideas to others. Students learn not only to interpret the language of math and science, but also to *use* that knowledge.

7. **Science and math are interdisciplinary.**
   Students work with the big ideas that connect subjects. Science and mathematics relate to many subjects. Science and technology are the obvious choices. Literature, music, art, social studies, physical education, and just about everything else, make use of science and mathematics in some way. If you want to understand what you are reading in the newspaper, for example, you need to be able to read the charts and the graphs taught in science and math classes.

ACTIVITIES THAT HELP STUDENTS UNDERSTAND SCIENCE AND MATHEMATICS

1. **Science and math are methods of thinking.**
   List all the situations outside of school in which you used science and math during the past week.
2. **Science and math both require knowledge of patterns and relationships.**
   Show all the ways fifteen objects can be sorted into four piles so that each pile has a different number of objects in it.
3. **Science and math are tools.**
   Solve these problems using the tools of math and science:
   • Will an orange sink or float in water?
   • What happens when the orange is peeled? Have groups do the experiment and explain their reasoning.
4. **Having fun, solving a puzzle with science and mathematics.**
   With a partner play a game of cribbage (a card game in which the object is to form combinations for points). Dominoes is another challenging game to play in groups.
5. **Using science and math as an art.**
   Have a small group of students design a fractal art picture.

6. **Applying the language of science and math.**
   Divide the class into small groups of four or five. Have the group brainstorm about what they would like to find out from other class members (favorite hobbies, TV programs, kinds of pets, and so forth). Once a topic is agreed on, have them organize and take a survey of all class members. When the data are gathered and compiled, have groups make a clear, descriptive graph that can be posted in the classroom.
7. **Designing interdisciplinary activities with science and math.**
   With a group design a song using rhythmic format that can be sung, chanted, or rapped. The lyrics can be written and musical notation added.

Like everyone else, a teacher's background is a little like a kaleidoscopic set of experiences and circumstances that have been influenced by education, time, and chance.

Like the all of us, it is a good idea for teachers to build on strengths and work on their weaknesses. A useful motto: Do your best today; do better tomorrow.

EXAMINING SCIENCE AND MATHEMATICAL PRINCIPLES

Science and mathematics education emphasizes problem solving, reasoning, teamwork, making connections, using technology, and understanding science–math relationships. As students learn good scientific reasoning skills, they have to figure out what the evidence tells them. The process involves being able to go beyond opinion and justifying their beliefs.

Although science and its math/tech tools sometimes miss their mark, they have a good track record when it comes to discovering empirical facts about the natural world. Yes, scientists sometimes make mistakes in their research and their conclusions. But that isn't a good reason to prefer opinion over facts. Skepticism is fine; purposeful falsification and ignorance aren't.

Digital technologies have a major role to play. Websites, search engines, data mining, social networks, and other Internet technologies are frequently part of today's science and math lessons (Andrews, 2012).

In the last century, teachers emphasized the memorization of facts and answering questions correctly. Now more attention is paid to helping students learn how science and mathematics relate to social problems, technology, creative innovation, and their personal lives (Peters & Stout, 2011).

STEM attends to content and the characteristics of effective instruction. Engaging students in active and interactive learning deepens their involvement in their academic work and their understanding of the subjects we teach. Good teachers know that students should have many opportunities to interpret science, technology, engineering, and math ideas and construct understandings for themselves.

The latest approaches in STEM teaching make an effort to be purposeful, provide meaningful activities with real applications that touch people's lives. By emphasizing collaborative inquiry, promoting curiosity, and valuing students' ideas, these subjects become more accessible and interesting.

A COLLABORATIVE MODEL FOR TEACHING STEM PROCESSES

Views of learning emphasize thinking processes within the learner and point toward changes that need to be made in the way that educators have traditionally thought about teaching, learning, and organizing the school classroom. Central to creating such a learning environment is the desire to help individuals acquire or construct knowledge.

That knowledge is to be shared or developed—rather than held by the authority. It holds teachers to a high standard, for they must have both subject matter knowledge and pedagogical knowledge based on an understanding of learning and child development.

The collaborative learning model for inquiry in science, technology, engineering, and math emphasizes the intrinsic benefits of learning rather than external rewards for academic performance. Lessons are introduced with statements concerning reasons for engaging in the learning task. Students are encouraged to assume responsibility for learning and evaluating their own work and the work of others. Interaction may include a discussion of the validity of explanations, the search for more information, the testing of various explanations, or consideration of the pros and cons of specific decisions.

The characteristics that distinguish new collaborative STEM learning revolve around group goals and the accompanying benefits of active group work. Instead of being told they need information, students learn to recognize when additional data are needed. They jointly seek it out, apply it, and see its power and value. In this way, the actual use of science, technology, engineering, and math information becomes the starting point rather than being viewed as an add-on. The teacher facilitates the process, instead of acting as a knowledge dispenser. Student success is measured by performance, work samples, projects, and applications.

Learning STEM in today's schools has a lot to do with exploring a problem, thinking, and posing a solution. This involves peers helping each other, self-evaluation, and group support for risk-taking. This also means accepting individual differences and having positive expectations for everyone in the group.

Making sure that there is plenty of time for active collaborative learning allows students to jointly address common topics at many levels of sophistication. This instructional method commonly involves having all students work on the same topic during a given unit. The work is divided into a number of investigatory or practical activities in which the students work and move from working alone to working in small groups.

OVERVIEW OF THE INTEGRATED STEM CURRICULUM

All students should:

- Understand numbers and operations, estimate and use computational tools effectively.
- Understand science and math subject matter including physical, life, and earth/space science, algebra, and geometry.
- Understand and use various patterns and relationships.
- Use observation and special reasoning to solve problems.
- Understand the themes and processes of science, technology, engineering, and mathematics.
- Understand and use systems of measurement.
- Become familiar with inquiry skills (pose questions, organize, and represent data).
- Focus on problem solving.
- Recognize reasoning and proof as essential and powerful parts of science and mathematics.
- Communicate ideas clearly to others by organizing and using thinking skills.
- Understand and make connections among the relationships of science, technology, engineering, and mathematics.
- Identify with the history, culture, and nature of STEM research.
- Understand and practice science, technology, engineering, and mathematics from a personal and social perspective.

These selected integrated standards are derived from the National Science Education Standards (National Academy Press, 2011), and the Curriculum and Evaluation Standards for School Mathematics (National Council of Teachers of Mathematics, 2000).

SAMPLE MATH ACTIVITIES

In an effort to link the integrated standards to classroom practice, a few sample activities are presented. The intent is not to prescribe an activity for a unique grade level; rather, it is useful to present activities that could be modified and used in many grades.

Activity One: Compare and Estimate

Objectives:

In grades K to 4, the curriculum should include estimation so students can:

- Explore estimation strategies.
- Recognize when an estimate is appropriate.
- Determine the reasonableness of results.
- Apply estimation in working with quantities, when using measurement, computation, and problem solving.

Science and math instruction in the primary grades tries to make classifying and using numerals essential parts of the classroom experience. Students need to go beyond counting and writing numerals to identifying quantities and seeing relationships between objects.

Directions:

1. Divide students into small groups of two or three students. Place a similar group of objects in a container which are color coded for each group. Pass out recording sheets divided into partitions with the color of the container in each box.
2. Have young students examine the container on their desks, estimate how many objects are present, discuss with their group, and write their guess next to the color on the sheet.
3. Next, have the group count the objects and write the number they counted next to the first number. Instruct the students to circle the greater number.
4. Switch cans or move to the next station and repeat the process. A variety of objects (small plastic cats, marbles, paper clips, colored shells, etc.) adds interest and is a real motivator.

Activity Two: Adding and Subtracting in Real-Life Situations

Objectives:

In the early grades, the science and mathematics curriculum should include concepts of addition and subtraction of whole numbers so that students can:

- Develop meaning for the operations by modeling and discussing a rich variety of problem situations.
- Relate the mathematical language and symbolism of operations toproblem situations and informal language.

When children are learning about the operations of addition and subtraction, it's helpful for them to make connections between these processes and the world around them. Story problems using ideas from science help them see the actions of joining and separating. Using manipulatives and sample word problems gives them experiences in joining sets and figure out the differences between them. By pretending and using concrete materials, learning becomes more meaningful.

Directions:

1. Divide students into small groups (two or three students).
2. Tell stories in which the children can pretend to be animals, plants, other children, or even space creatures.
3. Telling stories is enhanced by having children use unifix cubes or other manipulatives to represent the people, objects, or animals in the oral problems.
4. Have children work with construction paper or prepare counting boards on which trees, oceans, trails, houses, space stations, and other things have been drawn.

Activity Three: Solving Problems

Problem solving should be the starting place for developing understanding. Teachers should present word problems for children to discuss and find solutions working together, without the distraction of symbols. The following activities attempt to link word problems to meaningful situations.

Objectives:

Students will:

• Solve problems.
• Work in a group.
• Discuss and present their solutions.

Directions:

1. Divide students into small groups (two or three students).
2. Find a creative way to share 50¢ among four children. Explain your solution. Is it fair? How could you do it differently?
3. The children in your class counted and found there were 163 sheets of construction paper. They were given the problem of figuring out how many sheets each child would receive if they were divided evenly among them.

4. Encourage students to explain their reasoning to the class.
5. After discussing each problem, show the children the standard notation for representing division. Soon, you will find students will begin to use the standard symbols in their own writing.

Activity Four: Using Statistics: Supermarket Shopping

Statistics is the science or study of data. Statistical problems require collecting, sorting, representing, analyzing, and interpreting information.

Objectives:

Students will:

• Collect, organize, and describe data.
• Construct, read, and interpret displays of data.
• Formulate and solve problems that involve collecting and analyzing data.

Problem:

1. Your group has $2 to spend at the market. What will you purchase?
2. Have groups explain and write down their choices.
3. Next, have groups collect data from all the groups in the class.
4. Graph the class results.

Activity Five: Promoting Mathematical Modeling in STEM Activities

Modeling requires that students develop procedures to work on problem situations. Models explain or predict results, draw conclusions, and answer questions about real situations they encounter.

Fifth graders will explore a modeling experience by finding out the relationship among weight, distance, and balance using a lever. Students will be divided into groups. They will make sense of the problem by identifying, observing, and manipulating a mysterious lever. The mysterious problem includes four steps:

1. **Introduction**
   The teacher will show students a lever to examine without telling how it worked. A lever and some weights were provided for each group. Students will share ideas, take notes and describe how the tool would work.
2. **Exploration**
   Some students could try to manipulate the lever using one of its holes. After trying several holes, they may recognize that when they use the hole

in the middle, they achieve balance. By hanging weights on both sides of the bar, students will see how the lever worked. Hopefully, after several tries, they may realize that the weights they hung changed the balance. If they hang only one weight at a time from the middle hole, they can observe that balance is achieved. Students will be asked to record their observations on a recording sheet.

3. **Evidence**

   Students will be asked to organize their observations and provide a model based on the weights and distance on the lever to achieve balance. Students will work collaboratively together and record their findings.

4. **STEM Model Testing**

   Science, technology, engineering, and mathematics use a prediction process using models. The lever project is just one example. Students need to convert their experiences to other balance tasks and show it mathematically. One student revealed her thinking: "$2 \times 4 = 4 \times 2$." The numbers showed the weights on the lever.

Transdisciplinary STEM Activities

It is best to get away from the idea of splitting up the curriculum; instead, the new focus is looking at fusing various disciplines together. The next few activities try to accomplish this goal.

Activity Six: Creating Maps

Objectives:

Students will:

- Recognize, describe, extend, and create a wide variety of patterns.
- Represent and describe math and science relationships.
- Explore the use of variables and open sentences to express relationships.

As students measure many geometric figures, they uncover patterns and see relationships. After students have explored different shapes, the next part of the activity looks at patterns. The topic is called map making. A map is a pattern that we follow. It tells us where to go and how we go about getting there. The following activity has students follow a map, recording as they go.

Directions Map Exercise

1. Have children use graph paper and a pencil to draw a path. Each unit on the grid will represent one city block.
2. Write the labels for north, east, south, and west. Begin your map in the middle of the graph paper. Then follow this route.

A. Walk two blocks south.
B. Turn east and walk three blocks.
C. Turn south and walk one block.
D. Turn east and walk three blocks.
E. Turn south and walk four blocks.
F. Turn east and walk one block.
G. Turn south and walk three blocks
H. Turn west and walk half a block.

3. Compare your map to those of other students. How are they the same? How are they different? How would the map change if step B were a 90-degree turn west?
4. As a challenge problem: With your group, make a scale model of the local area. Include a "key" identifying symbols and directions.

Activity Seven: Estimate and Weigh Different Materials

Objectives:

Students will:

- Estimate the weight of different objects.
- Relate everyday experiences to the math/science measuring activity.
- Check their estimates.

Children will establish a link between their concrete everyday experience and their understanding of math and science abstractions through many different experiences in representing quantities and shapes. Representation helps children remember an experience and make sense by communicating it to others.

Directions: Measuring Activity

1. Fill several milk cartons with different materials such as rice, beans, clay, plaster of Paris, and wooden objects.
2. Seal the cartons and label them by color or letter. Tell the children what the materials are but do not identify the contents of a particular carton.
3. Have the children guess how to order the cartons by weight according to what they contain. Then, let the children order the cartons by weight, holding them in their hands and using the pan balance to check their estimates.

Activity Eight: Which Will Melt Quicker?

Objectives:

Students will:

Explore reasoning in science and mathematics.
Measure and compare quantities.
Write about their findings in their science and math journals.

Problem: Suppose you have a glass of water. It has the same temperature as the air. Would an ice cube melt faster in the water or air? Invite students to find out.

Materials: thermometer, water, ice cubes, two glasses (same kind), small plastic bag, salt, spoon.

Directions:

1. Fill one container with warm water and leave the other container empty.
2. Let the children see and feel the cups and the water. Explain: "We are going to put an ice cube in each container. Our problem: Which ice cube do you think will melt first?" Write the guesses on the board.
3. Have students measure the temperature inside the empty glass. Also, measure the temperature inside a glass of water. It should be about the same as the air temperature. If not, let the water stand a while.
4. Find two ice cubes of the same size.
5. Put one cube into the empty glass. Put the other in the glass of water.
6. Compare how fast the ice cubes melt.

Questions for Further Investigation:

1. How can you make an ice cube melt faster in water? Will stirring the water make a difference? Will it melt faster in warmer water? Does crushing the ice make a difference? Does changing the volume of the water matter?
2. How will ice cubes melt when other things are added to the water?
3. Will an ice cube melt faster in salt water? Does the amount of salt make a difference?
4. Students will measure and compare temperatures, hypothesize, experiment, and arrive at conclusions.

Evaluation: Have students explain their reasoning through writing about their experiment in their journals. Direct their discussion by asking them to explain what mathematics they used. What science skills were involved? What is the best way to show their data?

Activity Nine: Science and Math Metric Challenge

Objectives:

Students will:

* Solve problems working as a team.
* Explain directions.
    1. Divide the class into teams of three or four students.
    2. Give each team a list of challenges.

Team Challenges:

1. Find how many square meters of floor space each person has in your classroom.
2. If there are hundred students in the gym, how many cubic meters does each student have?
3. How many square meters does the school playground have?
4. Find the number of meters you must walk from our classroom to the principal's office and back.
5. Create a game with your group involving meters. Establish rules. Is luck involved? Write out the rules. Explain how to play the game to the class.

Activity Ten: Finding Your Heart Rate

Objectives:

Students will:

* Understand the heart and its system.
* Learn how to calculate their heart rate.
* Understand that heart rate changes depending on the kind of physical activity.
* Chart heart rate data.

Materials: classroom with room to move around, "calculate your heart rate" worksheet, pencils, timer, or clock.

Background Information: The heart is a pump. The heart pumps blood to different parts of our body. The number of times the heart pumps per minute is called heart rate. Heart rate changes when we do different activities.

Procedures: Introduce heart rate, have students guess how many beats per minute their heart is beating. Ask students to name activities they believe would change their heart rate. Then, ask if they think these activities would make the heart beat faster or slower than normal. Explain to students that there are two main areas in the body where it is easiest to find your heart rate (neck and wrist). Have students find their pulse. Make sure each student has found it. Practice counting while being timed for ten seconds.

Directions:

1. Pass out heart rate worksheet. Have students count heart rate while sitting.
2. Have students enter number on chart. Explain to students that this is how many times the heart beats in ten seconds.
3. Explain that heart rate is taken in one-minute intervals. Help students multiply their number by six in order to have the number of times their heart beats in one minute. Enter the number on the worksheet.
4. Do the same for activities of standing and running in place for thirty seconds.
5. After worksheet chart is filled in, ask students to write a sentence about a time when they felt their heart rate change.

Evaluation: Students' performance will be evaluated based on how much of the chart they are able to fill in and how well they participated in the activity.

Heart Rate Worksheet

Your heart rate is how many times your heart beats per minute.
Your heart rate changes.
Your heart rate is your pulse.
Your pulse is found in your neck and wrist.

Record your heart rate below: A useful instructional procedure: determine the purpose and scope of the lesson. Next, build on students' interests as you provide space for thinking, reflection, and discussion. Be prepared to push students' thinking forward with purposeful activities. Finally, share some of your lessons with other teachers and get suggestions for improvement.

Table 3.1

Record heart rate after...	Sitting	Standing	Running in place
10 seconds			
40 seconds			
60 seconds			

Whether it is individual or group work, it also makes sense for inquiry, scientific reasoning, and mathematical problem-solving skills to be integrated and utilized across the curriculum.

A SAMPLE OF ONLINE RESOURCES FOR STEM

Connected science, technology, engineering, and math activities and projects are part of today's science and mathematics curriculum. The learning process associated with STEM views science and math as underlying all engineering problems. Technology is viewed as an essential tool in the search for answers.

Organizations like the National Science Teachers Association (www.nsta. org) and the National Council of Teachers of Mathematics (www.nctm.org) can help. So can the following websites:

- AAAS Science NetLinks (http//sciencenetlinks.com).
  The American Association for the Advancement of Science has made available a wide range of lesson plans for just about every grade level.
- Discovery Education (www.discoveryeducation.com/teachers).
  Here, the content fits in nicely with state standards and there are links to a large number of teacher friendly math websites.
- PBS Teachers STEM Education Resource Center (www.pbs.org/teachers/stem and http://www.pbs.org/wgbh/nova/)
  Here, teachers and students can choose from thousands of resources, including lesson plans, videos, and interactive activities.
- NASA's Planet Quest Exoplanet Exploration (http://planetquest.jpl. nasa.gov)
  If students or teachers want to connect to experts, this is a good place to do it. Possibilities range from NASA's Jet Propulsion Laboratory (JPL) and images from space to the "Ask an Astronomer" podcast. Other possibilities: online games, activities, and submitting questions to experts.
- Learning and Teaching about the Environment (https://www.epa.gov/ students)
  Grades K to12 students and educators need access to quality homework resources, lesson plans and project ideas to learn and teach about the environment. Environmental education (EE) is a multidisciplinary approach to learning about environmental issues that enhances knowledge, builds critical thinking skills and helps students make informed and responsible decisions.
- Engineering: Go For It! (http://www.egfi-k12.org/)
  This website features a variety of tools to boost K–12 students' math and science abilities, enhance STEM teachers' instruction techniques, and

stimulate the classroom environment with great classroom activities, lesson plans, engineering projects, Web resources, and outreach programs.

- CODE.org (https://code.org/)

  Every student should have the opportunity to learn coding skills because information technology is one of the most critical technologies driving innovation in society. This wonderful site has plenty of ideas for teachers to introduce coding skills in the classroom at the earliest stage possible.

- Extreme Science (http://www.extremescience.com/)

  This website contains great information on the most unusual and extreme natural events that encompass earth science, the plant and animal kingdoms, weather, and much more hard-to-believe occurrences in nature. The site also offers complete collection of science and technology information students could readily use for their science and computer fair projects and class activities.

- NASA Education for Educators (http://www.nasa.gov/audience/foreducators/index.html)

  As a teacher, get ideas on how to inspire students to become future space explorers and astronauts!

- Science Buddies (http://www.sciencebuddies.org/)

  This is a great site to find creative science projects—special features are the family science activities and the use of Google classroom integration tools!

- Science Channel (http://www.sciencechannel.com/)

  This website offers a rich collection of science and technology videos that teachers can use in their classes. Feast also on a wide range of quizzes, games, and science news that can make any class sparkle!

- STEM Works! (http://stem-works.com/)

  This website is packed with quirky and funky activities students will enjoy—test students' skills with the reptile quiz, rescue an athlete in the Bionic Games, or simply follow the path of great whites with the Global Shark Tracker.

- Tech Rocket (https://www.techrocket.com/)

  This website has good information on how students aged ten years and above learn skills in coding, game design, and graphic design—very marketable skills in today's world.

- Tynker (https://www.tynker.com/)

  This website can help teachers get young students started with computer coding or programming. Tynker is a creative computing platform teachers can use to help students learn how to program and build games, apps, and more. The site offers self-paced online courses for children, so they can learn coding at home as well.

CONNECTING THE MATH STANDARDS TO THE CORE CONTENT STANDARDS

Standard 1: Understand Number and Operations
Students need to understand counting (represent one to one correspondence with concrete materials, match a set to a numeral).

Standard 2: Ability to Use Patterns, Algebra, Functions, and Variables
Students will understand and use functions (plus +, minus –, times ×, divide ÷).

Standard 3: Geometry
Apply geometry, understand shapes, use size, symmetry, congruence, and similarity.

Standard 4: Measurement
Use measurement to measure and compare lengths, widths; tell and write time in hours, half-hours; and use analog and digital clocks.

Standard 5: Data Analysis, Probability
Organize data and use charts, tables, graphs, and statistics to make sense out of mathematics.

Standard 6: Problem Solving
Find solutions, use strategies, take risks, make decisions, and get results.

Standard 7: Reasoning
Reasoning is connected to students' language development. Thinking and reasoning are important to math learning. Students should have experiences with deductive reasoning (moving from guesses to conclusions),

Learners should also be aware of inductive reasoning (informal reasoning using specific examples) and use evidence to make assumptions and form conclusions.

Standard 8: Communicating
This includes working in groups talking, listening, and expressing ideas. Students share information, explain ideas, and help each other.

Standard 9: Forming Conclusions
Many relationships are learned everyday by students making connections through their own experiences and applying math content to real-life situations.

Standard 10: Representing Math Relationships
Representing 0 (zero) is showing connections among math concepts improves understanding. When students are able to use different blocks, colored math

squares, and fraction pieces while performing math skills, learning becomes more enjoyable.

Teachers spend a lot of time trying out activities that they can use in their lesson plans. But little professional time is spent on STEM Instruction. The purpose of this chapter is to suggest some STEM lesson planning possibilities for science, technology, engineering and mathematics. We offer sample lesson possibilities that have been tested with teachers. The elements of these plans are constructed so that they can be changed to fit with any resource, program or science/math/technology/engineering school curriculum.

Lesson plans are outlines for designing lessons. The goal is to help teachers effectively plan and use long-range, short-range, or daily lesson plans.

Teachers begin by asking questions:

- What do I want students to learn?
- How will they learn it?
- How will I know when they have achieved it?

The setting of the classroom, class diversity, available materials, curriculum, and assessment methods is involved in planning lessons. The planning approach we suggest for activities and assessments help bring focus to quality science and math ideas, lessons, or units. Even though there are common elements in lesson plans, formats may vary as the lesson design changes.

A STEM lesson plan includes:

- *Objectives* or goals describing what the student should be able to do after the lesson.
- *Procedures* that state the instructional activities and ways that students learn the skills and concepts being taught.
- *Materials* that students, learning groups, and the teacher need to complete the lesson.
- *Assessment* or plan of how you will know if students have learned what you intended,
- *Accommodation* or way of adapting ideas, materials, and assessments so that students and groups are given different paths to learning.
- *Evaluation* that analyzes student performance, group collaboration, the success of the lesson, and teaching effectiveness.

Getting elementary and middle school students comfortable with STEM ideas and abstractions is a path to higher level education. Teachers may start a lesson by posing a thought-provoking question or problem. Struggling with the question/problem produces possible solutions. At the end of the lesson, groups can present their findings and discuss their solutions.

Improving the Traditional Teaching Model

> In a world that's changing at a dizzying speed, the intellectual and reasoning tools from the STEM subjects are critical. The power of the STEM subjects and the processes they possess can help us all approach the wide range of possibilities that occur daily.
>
> —Froschauer and Bigelow

Too Important to Be Left to Experts

Whether they are done individually or in groups, formally or informally, science/math/technology activities are simply too important to be left to the experts (Marshall, 2013).

SUMMARY, CONCLUSION, AND LOOKING AHEAD

In a world filled with the science- and math-based technological products, understanding these subjects is more important than ever. With the need for analytical skills and social intelligence goes any version of tomorrow. Still, when it comes to predicting the future, we are all *dancing in the dark*.

The best preparation is practicing as many steps as possible and getting ready for an uncertain and haphazard world. Also, serendipity plays an important role in inquiry and problem solving. But as the old saying goes: *if you don't have some idea where you are going, you are not going to get there.*

Both students and teachers have to prepare for the kind of real-world experience where flexibility and creativity are often the keys to success. Remember, unless you come up with new approaches, you'll never know what clever people might do with it.

When it comes to classroom assignments, it is often best to explain the *why* and *what* of a lesson. But if imaginative ideas and innovation are part of the plan, it is better not to explain *how* something should be done.

There are many ways to go about inquiry and problem solving. To move along any path to high achievement requires overcoming failure, so learn to deal with it. Preparation is the key, but winning and losing are dealt out by time, chance, and circumstance. The best approach is to simply do your best today and try to do better tomorrow.

Discoveries in the STEM subjects are built on replication, collaboration, and sorting out the real leads from the false ones. It's like exploring an unknown world, you never know what's just over the next hill. It could be a breakthrough; more likely, it's just another hill you have to climb over.

Sometimes, it takes years before the importance of a new finding really takes root and is recognized.

Most insightful new work is built on the foundation of prior knowledge. The same can be said for learning in general. When it comes to research in science and math, science and math experts will tell you that it is hard to sort out the mother from the mother-in-law of all discoveries.

An important educational goal is to ensure that the abilities of students are developed in ways that help them take hold of their futures. This requires knowledgeable and skilled teachers, students who are prepared and motivated, and parents who are supportive and informed.

The future is already here. It's just not evenly distributed yet.

—William Gibson

REFERENCES

Andrews, L. (2012). *I know who you are and I saw what you did.* New York, NY: Free Press.

Barnes, M. (2013). *Role reversal: Achieving uncommonly excellent results in the student-centered classroom.* Baltimore, MD: ASCD.

Bybee, R. (2013). *The case for STEM education: Challenges and opportunities.* Arlington, VA: NSTA Press (National Science Teachers Association).

Gibson, W. (2012). *Distrust that particular flavor.* New York, NY: G. P. Putnam's & Sons.

Leinwand, S. (2012). *Sensible mathematics: A guide for school leaders in the era of common core standards* (2nd edition). Portsmouth, NH: Heinemann.

Marshall, J. C. (2013) *Succeeding with inquiry in science and math classrooms.* Arlington, VA: National Science Teachers Association.

National Academy Press (2011). *National science education standards.* Washington, DC: National Academy Press.

National Research Council (NRC) (2011). *A framework for K–12 science education: Practices crosscutting concepts and core ideas.* Washington, DC: National Research Council.

Peters, J. & Stout, D. (2011). *Science in elementary education: Methods, concepts, and inquiries* (11th Edition). Boston, MA: Allyn & Bacon.

Tetlock, P. & Gardner, D. (2015). *Superforecasting: The art and science of prediction.* New York, NY: Signal, Penguin Random House.

Chapter Four

Engineering and Technology

Imaginatively Integrating STEM Subjects in the Classroom

Imagination is the source of every form of human achievement.

—Ken Robinson

Many issues surrounding the E and the T in STEM are overlooked because the emphasis has been on science and mathematics in many schools. Of course, many students use digital technology for many hours every day, but gaining an understanding of the underlying issues doesn't get much attention.

When engineering and technology topics are included in some school-based instruction, intellectual curiosity, creativity, and critical thinking are rarely part of the mix. The Internet is just one example of an engineering and technological tool that often gets a pass when it comes to criticism. The "Googling" pathway, for example, can even generate cynicism about really being informed.

We used to say that seeing is believing; now, if it can be found online, many of us believe it. The result: There are times when students are actually less informed because they think that they have access to all they need to know. We all need to realize that both the flow and the content of digital information are prone to manipulation. *Understanding* the data is more important than access to a glut of information.

With all our high-tech products, knowing and understanding can fall by the wayside (Lynch, 2016).

Learning-specific STEM skills matters, but so does developing broad-based analytic, problem-solving, and communication skills. This requires weaving competencies related to engineering and technology into the curriculum in ways that encourage creative thinking and collaboration.

Cultivating imaginative classroom work in STEM will not happen on a large scale unless professional systems of support are provided. As far as educators are concerned, a key element is providing professional development and group support in ways that emphasize teachers as learners and challenging them to improve their practices.

There are some who look at engineers and technologists as heads-down trench workers who are good at coding and using computers. Learning about technology and engineering is not just for students who want to become experts in these fields; however, it would be great if more students want to seriously study these fields. But even more important is helping *all students* become comfortable with these subjects so that they can become knowledgeable citizens who can function in today's world (National Research Council, 2011b).

ENGINEERING AND TECHNOLOGY: STEM POSSIBILITIES AND PROBLEMS

Like science and math, engineering and technology have a body of established knowledge that is vital for all citizens today. Like the subject of technology, the definition of engineering skills can be confusing. Some see these terms as a something of a vocational subject or a high-tech digital tool. Neither is true.

Any conversation about technology and engineering should consider the underlying issues, including both the possibilities and the problems. We should also consider the idea that both high-tech (like computers) or low-tech (like staplers or measuring tapes) can help solve some certain basic problems. More importantly, whatever we do requires that students be encouraged to think critically, write persuasively, and solve real-world problems creatively.

STEM instruction involves taking an interdisciplinary approach to teaching science, technology, engineering, and mathematics. The basic idea is to combine two or more STEM subjects in ways that open up possibilities for dealing with real-world problems. In this chapter, we focus on engineering and technology.

Forward momentum in any of the STEM subjects doesn't always have to be viewed as incremental successes, with occasional leaps forward. There are times when the change is quite dramatic.

One thing is for sure: maintaining privacy is increasing difficult with all cameras, sensors, and microphones that we have surrounded ourselves with. Worse yet, in the future, the technology will be even more efficient at worming its way into our private lives.

Uncertainty and curiosity underlie adapting to a constantly changing world. Even mistakes and dead ends can be turned to advantage and lead to breakthroughs. In fact, many imaginative discoveries arise out of unsuccessful attempts that stimulate conversation (Firestein, 2016).

STEM Guidelines

STEM is a way of organizing instruction. It integrates the disciplines of science, technology, engineering, and mathematics. Combining and creating traditional classroom units of study are also parts of the process.

1. *Concentrate on integrating the STEM disciplines.* Combining science, technology, and engineering with math, social studies, and art can help students make connections among ideas, and may forge innovative solutions to problems.
2. *Set up practical, personal, and social experiences.* Often, students don't see how learning relates to their lives. Is it interesting? Could this skill help me get a better job?
3. *Focus on tomorrow's skills.* Students need to know how they can access information they require to solve problems creatively and effectively. Collaboration and teamwork skills along with communication and critical thinking will be requisite skills of the twenty-first century (NSTA, 2009).
4. *Engineering and technology are not exclusively digital.* They can be as basic as a ball point pen or a squirt gun. Everyday items may be taken apart and studied (Moyer & Everett, 2012). Questions can be as simple as:
   • Why do squirt guns squirt?
   • How does the reservoir in the pen transfer ink to paper?
5. *Accelerate your students.* Teachers do this by challenging each student to figure out the best way to conduct their research, analyze data, and arrive at a conclusion. By having students apply what they have been learning in science, technology, engineering, and math, they will be able to engage in examining their data, evaluating results, and provide evidence.
6. *Solve problems when everything seems impossible.* Students need to work as a team to design an experiment, figure out how to do it, and present their concluding ideas.

Integrating STEM Plans and Principles

A plan can be thought of as a rule or principle that students put into practice. You can begin by concentrating on a few rules you can use in your classroom.
Why Technology and Engineering?

Engineering and technology are featured alongside the physical sciences, life sciences, and earth and space sciences for two critical reasons: To reflect the importance of understanding our world and to recognize the integrating of teaching and learning of science, engineering and technology.

—National Research Council

AN INTRODUCTION TO COMMON ENGINEERING

What do you think when you hear the word "engineering?" Everyone acknowledges that computers, airplanes, and genetically engineered plants are examples of technology, but for most of us, it stops there. Engineering is also about the very simple, common objects we use all the time. Scissors, can openers, zippers, etc. that can involve sophisticated engineering.

Many items that we take for granted in our daily activities have gone through many changes in order to meet human needs. Most things in our environment are artificial. These objects have been designed by people in order to solve a problem or provide a want or need.

Engineering Education

Engineering education is not a new idea. It's been around for over fifty years. What's new, however, is the inclusion of engineering in grades K through 12.

STEM Engineering Activities I

Children are all engineers. Primary teachers may not call it engineering but when children go about building a structure, that's what it is. If children have some background in at least several of the STEM subjects, ask them why they think that studying technology and engineering can be useful for everyone. They are enthusiastic about using creative design and trying to solve problems. Technologies are the ways we help change the world in which we live. People try to improve their surroundings to meet their needs and accomplish goals.

An important technology like transportation has lots of questions: Why are cars, ships, and airplanes needed? How do they help us? What problems do they create? How might they be improved to help solve these problems?

Floating into Understanding

A Classroom Example: Paper Airplanes

Before you start this lesson, you could give out an assignment for students to look up as many possibilities for paper airplanes as they can find on the Internet. This could be done at home or at school.

The students cheered as the paper airplanes crossed the finish line. They applauded as their group recorded the data and took notes. Another group had timed the event using calculators. These students explored airplane flight and how scientists investigate.

A STEM learning experience captures students' attention and motivates them to search and solve problems and to find out how scientists do their work. The paper airplane investigation connects to students' personal interests and experiences while cultivating enthusiasm. Students form ideas about forces, motion, and interactions. As a strategy for commanding interest and engagement, this instructional model (engage, explore, explain, elaborate, and evaluate) is helpful because it highlights the importance of exploratory opportunities (Bybee, 2013).

Engage

Students can display their knowledge and find out about the science behind plane flight. Groups are presented with a sheet of paper. They are to design and create a paper airplane. There are several types of planes to choose from—a glider, a normal paper airplane, or a dart advanced plane.

Students have to also predict how far their plane will fly. To do this, they need to first discover the forces that affect paper airplane flight. Below are scientific forces that act on paper airplanes:

1. Applied force
2. Normal force
3. Gravitational force
4. Air resistance

As a teacher, have students discuss what they think each one means and write down their ideas.

Next, have groups make a prediction. Predictions are formative assessments or guesses that help students with their designs. Students are not to test their ideas at this time.

Explore

Once students have worked together, they are ready to explore. Students decide on a way to test their investigation. Students plan and conduct an investigation that shows evidence on the effect of force on the motion of their plane. As students perform these activities, they practice the science and engineering common core state standards: asking questions, defining problems, investigating, and gathering evidence.

Students know the list of forces and now have a chance to try them on their plane. Most students may realize a paper airplane has an applied force where the student throws the paper airplane. During this lesson, some students may understand that the gravitational force pulls the plane down toward the ground. You could have students look at the concept of air resistance. The class will discuss the idea of lift forces exerted by the airplane on the surrounding air.

The next part of the activity will focus on testing. The group of three needs to have rules and assign certain group roles: Each group member is assigned a role: "thrower," "measurer," or "data recorder." Students rotate roles during each trial. When students are not conducting trials, they will cheer their other classmates on.

- One student shows where the plane lands.
- Another student measures the distance the plane flew—the distance flown is where the plane hits the ground.
- Redo flights when a plane's flight is blocked (runs into a table or wall).
- Throw plane from about 2.5 meters high.
- Each group conducts three trials.
- Do not disrupt other students' planes.
- Wear protective glasses when testing.
- When not participating in testing trials, sit by the gymnasium wall, until- testing is finished.

Explain

Finally, student groups have a chance to share their data and show the evidence they found. Every group makes a chart showing the airplane style and the distance it traveled. Students first make a prediction for each style and then record three trial distances.

Students work together to analyze data and provide feedback. Students will notice that the dart airplane flew further than the normal or glider planes. They explained why with the help of using force diagrams. Air resistance is greater for glide and normal planes than for dart models.

Table 4.1

	Airplane-Style Prediction: Distance Traveled in Feet				
Airplane style	Trial 1 (feet)	Trial 2 (feet)	Trial 3 (feet)	Total (feet)	Average (feet)
Normal					
Glider					
Dart					

Elaborate

Some students may want to test whether the size and types of paper will make a difference on the distance the plane traveled. The types of paper to be tested are construction paper, card stock, computer paper, plain white paper, and lined paper.

A student will introduce the idea of changing the weight by adding paper clips to the planes. Another student may suggest taking the airplanes outside to test the difference.

Evaluate

Like scientists, students will present their findings and will write a group lab report showing their conclusions.

Conclude

This investigation will hold students' attention and get students involved in discussing, thinking, and collecting data. They will communicate, compare experiences, and will find evidence to support their scientific claims. Through hands-on experiences, students will learn the necessary STEM scientific processes in exploring paper airplanes.

Standard 3 and 5 Motion and Stability: Forces and Interactions

Students will plan, investigate, gather data, and support arguments about gravitational forces. Science and engineering practices will include asking questions, defining problems, and carrying out investigations (Brown & Duduid, 2000).

STEM Wind Energy Engineering

Windmills are coming back in the form of wind energy. Old images of windmills are replaced by large-sized wind turbine farms seen today. While the design has changed, the basic idea remains the same. A blade captures the energy of the wind to turn a shaft that does work, like turning a coil of wire in a magnetic field to generate electricity. Wind is seen as a reliable source of energy for producing electricity.

In this lesson, students will build a simple pinwheel windmill. Test the power generated by different designs, record data, and compare the designs. Energy from the sun is changed into wind energy, which, then, becomes mechanical energy as the windmill turns.

Some history. People have always used energy of the wind for sailing as early as 5000 BC. Windmills were first used in China and the Middle East for pumping water and grinding grain around 200 BC. The technology made its way to America in the late nineteenth century. Windmills were used on farms and in rural areas to pump water and generate electricity. Today, wind energy is the fastest growing energy source and will be so for many years to come (U.S. Department of Energy [DOE], 2005).

Activity: Windmills Everywhere

Engage and review safety guidelines on the use of sharp objects like pins. Have students use safety goggles. To get students' attention, show them some windmill photos. These can be found by looking at search engines under the term "windmills." Using the Google search engine, for instance, select the "Google Images" option, and type in the word "windmills" in the search box to pull up many windmill images. Discuss the purposes of a windmill.

Students will be building a windmill. They will try to figure out how moving air causes the blades on the windmill to turn. Focus students' attention on the fact that windmills have many different designs and blades. Explain that they will be doing an experiment to see whether a three- or four-bladed windmill can lift a weight faster.

Explore: Each group of four students needs: two pencils with erasers, two straws, two file folders, two push pins, thread, paper clips, a ruler, scissors, tape, a stopwatch, a balance. One large box fan is needed for each group.

Pattern for Three- and Four-Bladed Pinwheels

1. Draw a square. Draw lines connecting the corners.
2. Draw a triangle. Draw a line from the tip of the top part of the triangle to the bottom of the triangle. Draw another line from the corner to the side of the triangle. The last line will be from the other corner to the side of the triangle.
3. Cut each pinwheel along the lines drawn. Note that each cut from the corners is about two-thirds of the way to the center.
4. Use pushpins to make holes at the dots in the corners and in the center.
5. Insert the pins through the corners and the center.
6. Cut six pieces of drinking straws. Slide them onto the pencil and glue all but two longer pieces, which will serve as bearings in which the pencil can spin freely.
7. Attach a piece of thread with a small piece of tape to the center of the pencil. A paperclip and two washers are attached to the other side of the thread.
8. Finally, push the pin and the pinwheel to the eraser with a small dab of glue.

Explain: Plan a way to test the pinwheel. Time how long it takes for the pinwheel to lift the weight. Organize your data and record. Share your results with the class. Compare. What is the evidence? Describe how the energy was transferred in the pinwheel your group made.

Extend: Observe what happened when a small motor was attached to a battery in the pinwheel.

Evaluate: Compare your group's results to those of the rest of the class. Determine the amount of power your design produced. Share with your classmates.

STEM Engineering Activities II

Keep in mind that technologies are the ways that people modify the world to meet human needs. Another engineering technology is electricity. Again, we start with students' questions: "Why do we need electricity?"

One way to generate students' responses is to have students make a list of all the ways they use electricity every day. Common student comments included "turning on the lights," "watching TV," "heating our house," and "cooking."

How is electricity created? Where does electricity come from? What are some of the problems associated with increased demands for electricity?

These questions provide many engaging student responses. The topic of electricity will stimulate students' past knowledge of scientific concepts and principles. Another important question today is: "What are some ways to reduce our need for electricity?" Or other questions might be: "From what types of sources can we generate electricity so that we don't harm the earth?" or "Why do we need to use renewable sources of energy?" "How are these different from nonrenewable sources of energy?" "Why might nonrenewable sources of energy be bad to use?" or "Why is carbon emission bad?" or "How could we generate electricity in ways to avoid problems of high carbon emission?"

Scientific investigations and facts are usually introduced to students by referring to phenomena and occurrences that they already know about from their environment. The goal is that children come to understand their observations and experiences from a scientific perspective, pose questions, and communicate and test simple ideas by planning experiments. Hopefully, students will work together to present their results and explain their thinking.

What is electricity? Dictionary describes it as a form of energy that occurs naturally (as in lightening) or is produced (as with a generator). It is expressed in terms of the movement and interaction of electrons (negatively charged elementary particles).

(Merriam-Webster Dictionary, 2004). This is a challenging topic for students. Electricity is a complex system. Students need to learn about power plants and how electricity works in their everyday lives.

DESIGNING COMMUNICATIONS TECHNOLOGIES

Ways of developing technologies used to create a human environment have often led to new ways of thinking and acting. New media have been disrupting the school curriculum for years. In the past, balanced solutions have been found. But digital devices and applications have been engineered in a way that makes them irresistible, unpredictable, and disruptive in both positive and negative ways.

Sometimes, new products or applications change everything; at other times, new tech tools don't work out at all. Available technologies have always rocked the social and classroom scene back-and-forth. The difference today is that the media and communication possibilities young people are dealing with have more power.

When it comes to the Internet, for example, it is important to learn how to sort out the bad from the good. If the information gathered is wrong—or if the knowledge transmission process is corrupt—the results will be dismal. The standard paradox of the last hundred years: our tools are developed faster and better than we have.

Whether it is in or out of the classroom, new ways of communicating and relating to information require a break from habit. There are times when the latest gadgets are so fascinating that reasonable approaches are overwhelmed. But remember, just because something is technically interesting and doable doesn't mean that it should be done. As any parent or teacher will tell you, just because children like something doesn't mean it's good for them.

Devices that keep us continually on course and fixed in time and place can destroy the magic of randomness. Also, the stupefying modern obsession with productivity can deny the whimsy that living fully demands. To avoid negative consequences, it is important to be aware of what is taking place in the world around you, become comfortable with thought-provoking ideas, and know how to sort through the glut of information.

With digital technology, questions must always be asked about the relationship between a problem and the dangers associated with how it gets solved. Has the Internet age made us lazy and forgetful, or has it "created a radical new style of human intelligence" (Thompson, 2013)?

Emerging electronic information and communications media can conjure up new environments for critical thinking, creativity, and teamwork. When used intelligently, they can help people do all kinds of things better. When done wrong, digital tools can destroy privacy, reputations, and a broad spectrum of basic rights. Self-discipline and a moral compass are needed to navigate around a wired world.

Although it can help, specific technical expertise is not always required for creative and innovative behavior in an increasingly digital century. There are times when all subjects need to be decoupled from technology so that students can master a topic by interacting with others to find openings to new ideas and inventions. Given the amount of time young people spend alone in front of a variety of screens, face-to-face human interaction is more important than ever.

To move in the direction of new opportunities requires more than preparing technicians. There are times when you need a plumber, but there are other times when you need a poet. At school, things like the humanities and whimsical thinking have roles to play as students strive to creatively master the STEM subjects.

Involving all students (equity) doesn't mean teaching to the lowest common denominator; excellence and acquiring a deep knowledge of a topic of subject matters even more. An important goal: engage young people in ways that will help them bring useful tech tools and imaginative thinking to whatever they do.

POSSIBILITIES AND PITFALLS IN A WIRED WORLD

We have been warned for decades that the United States is losing ground to international competitors because it does not provide its citizens with a competitive level of math, science, literacy, and technological skills.

The world of globalization and automation requires working harder, smarter, and frequent retooling. Literacy, math, and computer-related skills matter. But, so do critical thinking and problem solving in a technology rich environment. No matter what the subject, it is clear that America must upgrade the skills of its citizens or else, many will find the doors to the twenty-first century workplace closed to them (OECD, 2013).

The arrival of newly engineered technological possibilities has always been exhilarating, and frightening. And sometimes, it eventually becomes a routine part of life. A few decades into the twentieth century, the telephone went from being an oddity to being taken for granted. By the 1950s, television was rapidly expanding into being a commonly used device. Much like their parents' and grandparents' approach to new media in the past, the children and young adults of today see things like digital devices and the Internet as natural ways to extend their information and communication reach.

In a school setting, digital technologies can encourage analytical thinking and help students connect subjects to collaborative inquiry. But still, in some ways, the same technologies and their applications encourage uncritical users to know more-and-more about less-and-less. In fact, it is now possible for a person to travel through life or around the globe in a wired cocoon.

As educators increasingly take an active role in the development of educational technology, there is more of a reasoned curriculum connection. And the process itself can have something of a liberating effect on the imagination of all involved. Also, new technological possibilities can encourage new habits of the mind and fresh perspectives. To get all of this right requires everyone involved to view teachers' professional development as a necessity (rather than a luxury).

The same Internet that gives us more and more access to various viewpoints can also narrow our universe to well-worn grooves of redundant experiences. Unleashing the potential of digital media requires serious thinking, research, and experimentation in order to connect technology and the characteristics of effective instruction.

It is possible to use technology in a way that makes classroom learning more exciting and effective than ever. But like science, technology and learning can only progress by valuing thought, evidence, and usefulness over magical wonderment.

There is a lot of uncertainty out on the technological horizon. But when it comes to educating the young, teachers, technology, and the school curriculum will play major roles.

Technology is an important thing, but it's not the only thing. It is only helpful when it helps develop students' imagination, inventiveness, self-discipline, and ability to ask good questions.

Unexpected Possibilities in Engineering and Technology

Done wrong, activities and social connections that rely on digital technology can be sad and lonely. Among other things, learning to avoid the dark side of technology requires limiting the time one spends online and treating people online the way you want to be treated.

When done right (within a limited timeframe), digital technology can help an active meaning-centered curriculum to flourish in (and beyond) the classroom. But no matter what, the coming together of technologies like computers, video, satellites, and the internet is both evolutionary and revolutionary.

Social media and other online activities are not inherently asocial—assuming those using it have the discipline to put down the electronics. Children, for example, need to get by the dazzle and go outside to play. At school, authentic face-to-face communication and collaborative learning require carving out time for the devices to be disconnected.

Where do creative solutions come from and what sparks chains of creation? Computers and their associates are potentially powerful tools for communications, academic work, and innovation. Digital technology has been engineered to have the power to move literacy and learning patterns off established roads. By motivating students through the excitement of discovery, a wide assortment of technological tools can assist the imaginative spirit of inquiry and make lessons sparkle. It can even put students right out on the edge of discovery—where truth throws off its various disguises.

As human horizons shift, a sort of flexible drive and intent are required for innovation and progress. Technology can add power to what we do and help us kick against educational boundaries. Vivid images of electronic media can stimulate students as they move quickly through mountains of information, pulling out important concepts and following topics of interest.

The online process changes students' relationship to information by allowing them to personally shift the relationship of knowledge elements across time and space. Learners can follow a topic between subjects, reading something here, and viewing a video segment there. All of this changes how information is structured and how it is used. It also encourages students to take more responsibility for their own learning.

The negative side of online learning is the practice of paying partial attention. Students "no longer have time to reflect, contemplate, or make thoughtful decisions. Instead, they exist in a sense of constant crisis—on alert for a new contact or a bit of exciting news or information at any moment" (Rose, 2010).

All media as extensions of ourselves serve to provide new transforming vision and awareness.

—McLuhan and Lapham

Be ready for the unexpected. Things don't always go as planned. Alexander Graham Bell thought, when he invented the telephone, that it would be used to listen to distant symphony concerts. Thomas Edison thought that the phonograph record would be used to send messages. Some of the best thinkers often miss the potential of their inventions. For example, physicist Heinrich Hertz was the first to generate and detect radio waves, yet he dismissed the notion that his findings might ever have any practical value. Human advances often come from what may go unnoticed or seem trivial at the time.

Anything that changes perspective—from travel to technology—can help generate new ideas. The motivation is also there because it's usually more fun to do things where the unexpected may turn up than sticking with the easily predictable. Playfulness and experimentation can often open up creative possibilities, increasing the capacity to fashion ideas or products in a novel fashion. Creatively playing with various ideas, some of which may seem silly at the time, may result in getting lucky with one or two of them.

It is often difficult to detect the subtle happenstance and how we make room in our own lives for positive accidents to happen. Being exposed to different experiences and paying attention to what's going on in the world may help by opening all kinds of serendipitous possibilities. Training the eye to notice things goes a long way toward making unpredictable advances happen. Each new finding can open up fresh questions and possibilities—breaking the habits that get in the way of creative thinking and change. Rx for thinking in the future: follow your curiosity, leave doors open, use technological tools, and make room for good luck to happen.

PEDAGOGICAL OPPORTUNITIES WITH DIGITAL TOOLS

At school, the challenge is getting students to apply the same level of intensity to their schoolwork that they apply to social media and video games. Building a dynamic model of learning requires making good use of everything available. But action without vision can be a nightmare, and vision without action often leads nowhere. One thing is certain—new technology is bound to generate new ways of thinking, learning, and working (Levy, 2011).

The playful gleam in the eye is often an engine of progress. There are multiple tools and modes of expression that schools need to build on to promote the multitude of strengths and imaginations found in all students. Many schools are mired in unproductive routines that prevent teachers from making creative breakthroughs.

Educators are not often able to take enough time to go back to the drawing board and use good data and experience to get it right. Reflection and changing approaches take time, space, support, and time for collegial professional development to make all the changes required.

Powerful forms of face-to-face learning within schools must not be neglected as we sort out the new media possibilities. Intelligent use of electronic forms of learning has proven to be helpful in improving student learning. Electronic learning, at least, presents the possibility of making a contribution. And it is a useful supplement to the professional development toolkit.

It is difficult to unravel issues of creativity or analyzing without taking into account influential media, like computers and television. They have a tremendous impact on children. We shape them and they shape us. Some are often written about as Lady Caroline wrote about Lord Byron—"mad, bad and dangerous to know."

Some see technology, such as computers and the Internet, as an example of engineering producing a particularly dangerous enemy. Among other terrible things, the dark side of high-tech includes creating a culture of electronic peeping Toms without a moral foundation.

Creativity can fall by the wayside if mobile devices are allowed to be a source of distraction in the classroom. It is so bad that many schools (like some concert halls, restaurants, and airlines) have installed devices on premise that prevent any texting, outside calls, or other intrusions during certain hours.

A few schools totally ban all devices on school property. Oddly enough, the expensive Walden private schools that enroll the children of many Silicon Valley parents ban the presence of digital devices at many of their elementary schools.

There may be times that a connection to the Internet can expand the imagination by opening up the possibility of discovering new things and support knowledge acquisition. But with or without tech companions, openness to experience helps generate a creative drive for exploration (Kaufman & Gregoire, 2016).

To be valuable, educational technology must be engineered and used in a manner that contributes to the improvement of learning. Digital devices and their accessories should be engineered (designed) to help open doors to reality and provide a setting for reflection. By making important points that might otherwise go unnoticed, these technological tools can help students refine and use knowledge more effectively. For example, computers can use mathematical rules to simulate and synthesize life-like behavior of cells growing and dividing. It's a very convincing way to bring the schooling process to life.

The yeast of knowledge, openness, and enterprise raises the need for a multiplicity of learning media and technological tools. Schools can teach students to recognize how technology can undermine social values, human goals, and national intention. They can also help students learn to harness these powerful tools so that they might strengthen and support the best in human endeavors.

It is our belief that when the pedagogical piece is in place, technology can support and strengthen the best approaches to student learning. This can change not only what teachers teach, but also how they go about doing it.

As new technologies and related products start to fulfill their promise, students will become active participants in knowledge construction across a variety of disciplines. We have had only a glimpse of the technological gateways to learning that will open in the twenty-first century. As state-of-the-art pedagogy is connected with state-of-the-art technological tools, the way knowledge is constructed, stored, and learned will be fundamentally altered.

THE FUTURE OF NARRATIVE IN CYBERSPACE

The idea of print as an immutable cannon may or may not be a historical illusion. One thing for sure, the way print is being mixed electronically with other media changes things. Although the American book industry is rushing into the emerging electronic literary market, book pages made of paper won't go away. Amazon's Kindle is but one example of an electronic reader.

One way or another print is here to stay. Even the doomsayers usually use books to put forward their argument that the medium is a doomed and outdated technology. In the future, will books be confined to dusty museum libraries? No, they will remain an elegant, user-friendly medium. With printed books, you don't need batteries and don't have to worry about the technological platform becoming obsolete and unusable.

There are at least two fairly new digital approaches to books that are finding a niche in the literacy universe. One is much like an electronic version of printed books. The other approach to electronic books is interactive and visually intensive. It takes the narrative and places it in randomly accessible blocks of text, graphics, and moving video. With some e-book stories, students must learn to go beyond merely following the action of the plot to learn about characters, explore different ideas, and enter other minds.

An interactive e-book story places students in charge of how things develop and how they turn out. Participants are able to change the sequence or make up a new beginning to a multidimensional story. They can slow up to find out additional information and they can change the ending. Navigating interactive stories with no fixed center, beginning, or end can be very disconcerting to the uninitiated. It requires a set of different "reading" skills.

To make sense of the anarchy and chaos, a reader has to become a creator. This means following links around so that they can discover different themes, concepts, and outcomes. "Interactive Storytime" is an early example. It tells stories with narration, print, music, sound effects, and graphics. Children can click on any object and connect spelling to the pronunciation.

Literature has traditionally had a linear progression worked out in advance by the author. The reader brought background knowledge and a unique interpretation to what the author had written. But it was the author who provided the basic sequential structure that pushed all readers in the same direction. Computer-based multidimensional literature is quite different. The reader shapes the story line by choosing the next expository sequence from a number of possibilities.

With early versions of interactive computer-based literature, readers are connected to a vast web of printed text, sound, graphics, and life-like video. When key words or images are highlighted on a computer screen, the reader clicks what they want next with a mouse (or finger on a touch screen) and the reader hops into a new place in the story, causing different outcomes. With a virtual reality format, the "reader" uses their whole body to interact with the story. Whatever the configuration, interactive literature causes the user to break down some of the walls that usually separate the reader from what's being read.

The forking paths in this electronic literature pose new problems for readers—like how do you know when you have finished reading when you can just keep going all over the place. Judy Malloy's *Its Name Was Penelope*, for example shuffles four hundred pages of a fictional woman's memories so that they come together very differently every time you read it. However, nowadays, the ambiguity of these programs isn't as bothersome as it used to be.

Today, children are used to television and computer programs that deal with quick movement between short segments of information. In addition, many video games and computer simulations require students to wander in a maze of ideas. As a result, children are usually not as disoriented by the various forms of interactive literature as adults.

Odd varieties of e-stories can be found, free of charge, on the Internet. Some are free, some are comparable to a traditional book in price, and others require a multimedia or virtual reality platform. Many of these efforts at constructing interactive stories were more like interactive comic books than literature. Programs are becoming more sophisticated and are giving us some advance warning of a new literary genre. One thing is for sure—something important to the future of curriculum and instruction is happening.

Online Peer Tutoring

Whether it is offline or online, peer tutoring involves several methods. In cross-age tutoring, older students tutor younger students. In cross-ability tutoring, a student who has a good grasp of the subject can tutor a struggling student. Reciprocal tutoring involves a structure for interaction that is tied to specific academic goals.

Conveying Meaning with Powerful Visual Models

One way to enhance the power and permanency of what we learn is to use visually based mental models in conjunction with the printed word. Inferences drawn from visually intensive media can lead to more profound thinking. In fact, children often rely on their perceptual (visual) learning even if their conceptual knowledge contradicts it. In other words, even when what's being presented runs contrary to verbal explanations, potent visual experiences that can push viewers to accept what is presented can be used.

Children can become adept at extracting meaning from the conventions of video, film, or animation—zooms, pans, tilts, fade outs, and flashbacks. But distinguishing fact from fiction is more difficult.

The ability to understand what's being presented visually is becoming ever more central to learning and to our society. Most of the time, children construct meaning for television, film, computer, or Internet content without even thinking about it. They may not be critical consumers, but they attend to stimuli and extract meaning from subtle messages.

The underlying message children often get from the mass media is that viewers should consume as much as possible while changing as little as possible. How well content is understood varies according to similarities between the viewers and the content. Viewers' needs, interests, and age are also other important factors in how content is processed. Sorting through the themes of mental conservatism and material addition requires carefully developed thinking skills.

Meaning in any medium is constructed by each participant at several levels. For better or for worse, broadcast television used to provide us with a common culture. When viewers share a common visual culture, they must also share a similar set of tools and processes for interpreting these signals (construction of meaning, information processing, interpretation, and evaluation).

The greater the experiential background in the culture being represented, the greater the understanding. The ability to make subtle judgments about what is going on in any medium is a developmental outcome that proceeds from stage to stage with an accumulation of experience.

Equity doesn't mean designing lessons for the lowest common denominator. Equity and excellence are not mutually exclusive goals.

Relying upon a host of cognitive inputs, individuals select and interpret the raw data of experience to produce a personal understanding of reality. What is understood while viewing depends on the interplay of images and social conditions.

Physical stimuli, human psychology, and information processing schemes taught by his or her culture helps each person make sense of the world. In this respect, reacting to the content of an electronic medium is no different from any other experience in life. It is just as possible to internalize ideas from electronic visual imagery as it is from conversation, print, or personal experience. It's just that comprehension occurs differently.

Reflective thought and imaginative active play important parts in the growth process of a child. Even with a "lean back" passive medium like television, children must do active work as they watch, make sense of its contents, and utilize its messages. With a "lean forward" medium like the

Internet, this work is fairly evident. Evaluative activities include judging and assigning worth, assessing what is admired, and deciding what positive and negative impressions should be assigned to the content. In this sense, children are active participants in determining meaning in any medium.

Adults Influence How Children Learn to Assess Media Messages

Although children learn best if they take an active role in their own learning, parents, teachers, and other adults are major influences. They can significantly affect what information children gather from television, film, or the Internet. Whatever the age, critical users of media should be able to:

- Understand the grammar and syntax of a medium as expressed in different program forms.
- Analyze the pervasive appeals of advertising.
- Compare similar presentations or those with similar presentations or those with similar purposes in different media.
- Identify values in language, characterization, conflict resolution, and sound/visual images.
- Utilize strategies for the management of duration of viewing and program choices.
- Identify elements in dramatic presentations associated with the concepts of plot, storyline, theme, characterization, motivation, program format, and production values.

Parents and teachers can affect children's interest in media messages and help them learn how to process information. Good modeling behavior, explaining content, and showing how the content relates to student interests are just a few examples of how adults can provide positive viewing motivation. Adults can also exhibit an informed response by pointing out misleading messages, without building curiosity for undesirable programs.

The viewing, computer, and Internet using habits of families play a large role in determining how children approach a medium. The length of time parents spend watching television, the kinds of programs viewed, and the reactions of parents and siblings toward programming messages all have a major influence on a child. If adults read and there are books, magazines, and newspapers around the house, children will pay more attention to print. Influencing what children view on television or the Internet may be done with rules about what may or may not be watched, interactions with children during viewing, and the modeling of certain content choices. A tip for parents: it's usually a good idea to keep the computer in the family room.

Whether coviewing or not, the viewing choices of adults in a child's life (parents, teachers, etc.) set an example for children. If, for example, parents are heavy watchers of public television or news programming, then, children are more likely to respond favorably to this content. If they stop their conversation in mid-sentence to answer their iPhone, it sends a signal to children about what's most important.

If parents make informed intelligent use of their TV viewing and gadgets, then, children are likely to build on that model. Clearly, influencing the settings in which children watch TV, use devices, or work on a computer is a major factor. Turning the TV set off—or putting the iPhones in airline mode—during meals sets a family priority.

Parents can also seek a more open and equal approach to choosing television shows and using other media use by interacting before, during, and after the program or app is finished.

When it comes to the Internet, we suggest keeping a child's use of their computer or iPhone in the family room—or at least where adults can observe what's going on. Using their devices in private (isolated) space is a bad idea. Time limits must be in place for electronic gadgets. Parents can even collect the iPhones and organize formal or informal group activities outside the house that provide alternatives.

The power of new technology is a major willpower challenge for everyone. Teachers are key players in solving the problem of technology distraction and "addiction." But it is increasingly clear that the education of children is a shared responsibility. The tech and media companies are part of the mix. Also, everybody, including parents, can play a central role especially if they have connections with what's going on in the schools.

By working together, parents and teachers can use the television, computers, iPhones, and the Internet to encourage students to become more intelligent media consumers. When it comes to schooling, it is the teachers who will be the ones called upon to make the educational connections entwining varieties of print and visual media with the basic curriculum.

Communications Technology and Public Conversations

A democratic community is defined by the quality of its educational institutions and its public conversations. Democracy often becomes what it pays attention to. American national values, supported by our constitution, require an educated citizenry that can think, respond to leaders, and are willing to actively go beyond the obvious. Patriotism isn't just the flag and stern rhetoric, it's a thinking, decent, and literate society. Exercising citizenship in a world of accelerated change requires the preservation of our human values.

Ignoring the societal implications of technology means ignoring looming changes. Whether it's technologically induced passivity or the seductive charms of believing in simplistic technological solutions, it is only through the educational process that people can gain a heightened awareness of bright human and technological possibilities. The question that we need to answer is: how might the technology be used to spark a renaissance in human learning and communication?

The long-term implications of recent changes in information and communications technology are important, if not frightening. The convergence of technologies is causing a major change in societal behaviors, lifestyles, and thinking patterns. With few people monitoring digital technology or theorizing its health, the human race is being forced to swim in an electronic sea of information and ideas. In today's world, there is little question that reality is being shaped by electronic information and electronic illusions.

Activities for Making Sense of Visual Media

1. **Help students critically view what they see.**
   Decoding visual stimuli and learning from visual images require practice. Seeing an image does not automatically ensure learning from it. Students must be guided in decoding and looking critically at what they view. One technique is to have students "read" the image on various levels. Students identify individual elements and classify them into various categories and then relate the whole to their own experiences, drawing inferences, and creating new conceptualizations from what they have learned. Encourage students to look at the plot and story line. Identify the message of the program. What symbols (camera techniques, motion sequences, setting, lighting, etc.) does the program use to make its message? What does the director do to arouse packaging, color, and images) that influence consumers and often distort reality? Analyze and discuss commercials in different media. How many minutes of ads appear in an hour? How many ads do you have to sort through before you can watch a program or use a search engine get to some websites? What should be done about the ad glut?

2. **Create a scrapbook of media clippings.**
   Have students keep a scrapbook of newspaper and magazine clippings on computers, the Internet, Facebook, Google, television, and some of the other inhabitants of cyberspace. Newspapers and magazines are good sources of articles; *The New York Times* Science section is a good source for upper grade students. Ask students to paraphrase, draw a picture, or map out a personal interpretation of an interesting technology article. Share these with other students.

3. **Create new images from the old.**
   Have students take rather mundane photographs and multiply the image, or combine it with others, in a way that makes them interesting. Through the act of observing, it is possible to build a common body of experiences, humor, feeling, and originality. And through collaborative efforts, students can expand on ideas and make the group process come alive.
4. **Role play communicating with extraterrestrial life.**
   Directions for students: If extraterrestrial life has already contacted us, think about how to respond. Think creatively and scientifically about how you would explain this planet to some inquisitive aliens. Explain human life and media devices; try to view our planet as a whole.
5. **Use debate for critical thought.**
   Debating is a communication model that can serve as a lively facilitator for concept building. Take a current and relevant topic, and formally debate it online or face to face. This can serve as an important speech and language extension. For example, the class can discuss how mass media can support everything from commercialism to public conformity and the technological control of society. The discussion can serve as a blend of technology, social studies, science, and the humanities.

   Electronic media and social patterns are constantly shifting through various stages of acceptance and use. Communications now come in many forms and from many locations, served up on a variety of digital devices. So, it is little wonder that an important instructional goal is giving young people the skills needed to swim through today's technological crosscurrents.

DIFFERENT MEDIA SYMBOL SYSTEMS

Print and visually intensive media take different approaches to communicating meaning. Print relies upon the reader's ability to interpret abstract symbols. A video or computer screen is more direct. Whatever the medium, thinking and learning are based on internal symbolic representations and the mental interpretation of those symbols. When they are used in combination, one medium can amplify another.

We live in a complex society dependent on rapid communications and information access. Life-like visual symbol systems are comprised in part, of story structure, pace, sound track, color, and conceptual difficulty. Computers, the Internet, television, and digital devices are rapidly becoming our dominant cultural tools for selecting, gathering, storing, and conveying knowledge in representational forms.

Various electronic symbol systems play a central role in modern communications. It is important that students begin to develop the skills necessary for interpreting and processing the full range of media messages. Symbolically different presentations of media vary as to the mental skills of processing they require. Each individual learns to use a media's symbolic forms for purposes of internal representation.

To even begin to read for example, a child needs to know something about thought–symbol relationships. To move beneath the surface of electronic imagery requires some of the same understandings. It takes skill to break free from a wash of images and electronically induced visual quicksand. These skills don't just develop naturally; training is required to develop critical media consumers who are literate in interpreting and processing print or visual images.

Unlike direct experience, print or visual representation is always coded within a symbol system. Learning to understand that system cultivates the mental skills necessary for gathering and assimilating internal representations. Each communications and information medium makes use of its own distinctive technology for gathering, encoding, sorting, and conveying its contents associated with different situations. The technological mode of a medium affects the interaction with its users, just as the method for transmitting content affects the knowledge acquired.

The closer the match between the way information is presented and the way it can be mentally represented, the easier it is to learn. Better communication means easier processing and more transfer. At its best, a medium gets out of your way and lets you get directly at the issues. New educational choices are being laid open by electronic technologies. Understanding and employing these technological forces require interpreting new media possibilities from a unique and critical perspective.

Understanding and Creating Electronic Messages

Understanding media conventions helps cultivate mental tools of thought. In any medium, this allows the viewer new ways of handling and exploring the world. The ability to interpret the action and messages requires going beyond the surface to understanding the deep structure of the medium. Understanding the practical and philosophical nuances of a medium moves its consumers in the direction of mastery.

Simply seeing an image does not have much to do with learning from it. The levels of knowledge and skills that children bring with them to the viewing situation determine the areas of knowledge and skills development acquired. Just as with reading print, decoding visually intensive stimuli and learning from visual images require practice.

Students can be guided in decoding and looking critically at what they view. One technique is to have students "read" the image on various levels. Students identify individual elements, classify them into various categories, and then, relate the whole to their own experiences. They can, then, draw inferences and create new conceptualizations from what they have learned.

Planning, visualizing, and developing a production allow students to critically sort out and use media techniques to relay meaning. Young producers should be encouraged to open their eyes to the world and visually experience what's out there. By realizing their ideas through media production, students learn to redefine space and time as they use media attributes such as structure, sound, color, pacing, and imaging.

The field of technology and its educational associates are in a period of introspection, self-doubt, and great expectations. In a world, awash with different types of educational media, theoretical guidelines are needed as much as specific instructional methods. It is dangerous to function in a theoretical- or research-deprived vacuum because rituals can spring up that are worse than those drained away. As schools are faced with aggressive marketing for electronic devices, we must be sure that a pedagogical plan that incorporates technology is in place.

For technological tools to reach their promise requires close connections among educational research, theory, and classroom practice. Across the curriculum, new standards place a high premium on creative and critical thinking. So, it is clear that curriculum and professional development will pay even closer attention to such skills.

Reaching students requires opening students' eyes to things they might not have thought of on their own. This means using technological tools as capable collaborators for tapping into real experience, fantasies, and personal visions. This way, previously obscure concepts can become comprehensible, with greater depth, at an earlier age.

Technology and metacognitive strategies can come together as students search for data, solve problems, and graphically simulate their way through multiple levels of abstraction. The combination of thoughtful strategies and the enabling features of media tools can achieve more lasting cognitive change and improved performance.

LEVELING THE LEARNING FIELD IN DIGITAL SPACE

Historically, Americans hadn't needed a rigorous education. Wealth had made rigor optional.

—Paraphrasing Amanda Ripley

Equity doesn't mean teaching to the lowest common denominator. For high-quality educational experiences, equal educational opportunities must go hand in hand with academic excellence.

As we put together the technological components that provide access to a truly individualized set of learning experiences, it is important to develop a modern philosophy of teaching, academic standards, learning, and social equity.

Harmonizing students' present with the future requires more than reinventing the schools. Many children are behind on their first day of school, so efforts to improve schooling must extend before and beyond the classroom door. For all learners to thrive academically, it helps to have the benefits of preschool, high-quality teachers, health care, and engaged family support.

When it comes helping all students reach their educational potential, it takes a lot more than technology. But tech tools can lend a hand in the effort to devolve the corrosive brew of poverty and neglect that eats away at the fabric of democratic life.

While new information and communications technologies have the potential to help make society more equal, it sometimes has the opposite effect. At home and at school, equal access to new technological worlds is a long way from reality.

Many school districts lack the money to train teachers to use digital technology effectively. Searching for information on the Internet is common practice in many schools. But even in the schools that have plenty of computers, the focus is on typing, drill, and workbook-like practice.

Affluent schools are more likely encourage students to use digital technology for creative exploration—like designing multimedia presentations and collaborating with classmates in problem-solving experimentation (Ripley, 2013).

Everyone deserves access to the tools associated with a provocative and challenging curriculum. To become an equal instrument of educational reform, digital technology must do more than reinforce a two-tier system of education. Otherwise, many children will face a discouraging picture of technological inequality.

There are serious social consequences surrounding the inequalities surrounding the use of the latest telecommunications and information technologies. Connections with other students, databases, and library resources have the potential to change the way that information and knowledge is created, accessed, and transmitted. Those denied informed access may very well find their ambitions stifled as they fall further and further behind the more fortunate.

The challenge is to make sure that this information is available for all in a twenty-first century version of the public library.

Digital technology gives us the ability to change the tone and priorities of gathering information and learning in a democratic society. Taking the right path with this tool requires learning to use what's available today and building a social and educational infrastructure that can travel the knowledge highways of the future.

Thinking: Slow, Accelerating, and Fast

Electronically connecting the human mind to people and global information resources may shift human consciousness in ways similar to what occurred in moving from an oral to a written culture. The ultimate consequences are unclear. But the development of basic skills, habits of the mind, wisdom, and traits of character will be increasingly affected by the technology.

Computer-based simulations can simulate invisible things like molecular reactions and static electricity. Also, with role-playing interactive activities from around the world, it is possible to foster thinking skills and collaboration.

The instructional activities that are most effective and popular are those that provide for social interaction and problem solving. Information can be embedded in visual narratives to create contexts that give meaning to dry facts.

Interactive digital technology can challenge students on many levels and even serve as a training ground for responsibility, persistence, and collaborative inquiry. It's relatively easy to buy the hardware and get children interested. The difficult thing is finding a way to connect to deep learning in a manner that advances curriculum goals.

New technology is a double-edged sword. Laptops can be used to record or take notes on the day's activities and presentations. They can also be a distraction as students check e-mail, Facebook; a few even use the technology to cheat during class. Cell phones can also be used to gather educational useful information and record important parts of a class. But they can also be brazenly used to send instant messages, waste time texting friends, and disengage in class altogether.

To stay on a useful curriculum and instruction path at school, it is important to recognize the fact that many students today have been brought up in an age of whiz-bang gadgets. Their approach to media and daily life is one of constant contact and multitasking. Technological expertise and curiosity are one thing; constant distraction quite another.

On the first day of class, teachers need to set and post specific rules and consequences for off-task activities. Decide if or when cell phones or laptops can be used during class. Remember to tell students if you don't want a lesson to be recorded. Do not leave it open for debate; enforce the posted rules consistently and fairly.

Schools need more adults than ever: teachers' aides, parents, older students, and more. Technological tools can get in the way of learning. But on the positive side, they can be a unique and useful supplement that allows teachers to enhance their lessons and try out some new things.

Generating Questions about Technological Surveillance

When a new medium emerges, it generates new ways of thinking and the opening of new realities. Asking probing questions about such things as the intrusion of surveillance technology and the collection of communications is part of what it means to be technologically literate today.

Internet users may know that the websites they use often work with online advertisers to target products and services that might interest you. What isn't as well known is that these companies can track you as you travel around and move among devices like computers, tablets, and cell phones. Also, they make up profiles without an individual's permission that can be accessed by employers, governmental agencies, and third party marketing firms.

Many countries use surveillance technologies, but the United States has them all beat. The National Security Administration (NSA), for example, has a huge collection of American communications and a record of just about every cross-border interaction. To make things easier for themselves, the NSA secretly inserts "back doors" into high-tech products and international encryption systems. Also, they collaborate with big software companies to have special access routes built into their systems.

Confidential information on cloud storage is wide open to some businesses and governmental agencies. By the way, "cloud storage" means that organizations can now purchase services from third party information technology vendors who have the hardware, software, and services needed to store a customer firm or individual's data. All customers have to do is pay a minimum fee based on usage for the data storage provided by these "cloud service providers." With cloud storage, for instance, it has been easy for a number of foreign and domestic agencies have been caught spying on love interests and the personal lives of celebrities. Others—from foreign spy agencies to individual hackers—sometimes play the same game with personal and commercial data.

Are we doing enough to prepare for the level of surveillance possible today?

The public is just beginning to learn about computerized tools like the Biometric Optical Surveillance System operated by the Department of Homeland Security. The program instantly connects to everything from live video cameras to stored driver's license photos, Facebook pages, global positioning system (GPS) signals, and Google searches. The result: individual faces are can now be identified within shorter periods of time using massive amounts of data, including what a person is up to at any given moment.

The National Security Agency (NSA), for example, says that they can collect whatever they want from anyone, anytime, anywhere. The overall goal seems a desire to have access to virtually everything available in the digital world. Encryption helps, but it just slows them down as they track anyone who uses a computer, mobile device, or router.

Beyond the gathering of personal information, the NSA intends to influence and shape the trends in high-tech industries in ways that are not yet known to users. Other agencies, like the Defense Department, are also looking for help from Silicon Valley. Artificial intelligence (AI), virtual reality, robotics, and cybersecurity are just a partial list at the intersection of government and business interests.

The end result of any government gathering and storing of every conceivable type of personal pictures, conversations, and messages is bound to have a negative effect on the open exchange of ideas. This is more likely to happen when new algorithms used go beyond facts and concepts to interpretation and the labeling of individuals and groups. The whole thing seems to invite mission creep and abuse.

The pros and cons of business/government surveillance make for a good small group and full class discussion. Legal or illegal, data miners and advertisers have the advantage over consumers.

LEARNING TO EVALUATE ONLINE
ACTIVITIES AND SOFTWARE

Curriculum is a cooperative and interactive venture between students and teachers. On both individual and social levels, those directly involved with the process must be taken seriously. Working together, they can decide what benefits are gained from particular software programs. After all, the software user is in the best position to decide if the program is taking people out of the process—or whether learners are in control of the computers. Good software programs let students learn together at their own pace—visualizing, talking together, and explaining abstract concepts so that they can relate them to real-life situations.

There are time-consuming evaluation issues surrounding the multitude of software programs to be dealt with. Multimedia, simulation, virtual reality, microworlds, word processing, interactive literature, spreadsheets, database managers, expert (AI) systems, or getting the computer in contact with the outside world all increase the potential for influencing impressionable minds. That is too large a universe for the teacher to figure out alone. Teachers can help each other. Another thing that helps the task become more approachable is getting the students to take on some of the responsibility.

Children can learn to critique everything from computer software to Internet websites much as they learned to critique the dominant media of yesteryear—books. One of the first tenants of book review criticism is to critique what is usually taken for granted. Also, having at least a little affection for what is being reviewed helps. Students can do book reports, review computer software, and discuss the quality of sites on the Internet. It helps if they sift through some of the software reviews in newspapers, magazines, and journals—online and offline.

Examining online activities and software: students can quickly check to see if the flow of a program makes sense. Next, they can try out the software as they think a successful or unsuccessful student might. Is the software easy to learn to use? Are there handy help guides when problems occur? What happens when mistakes are made? How are the graphics? Do you think you can learn anything from this? Is it exciting to use?

Without question, teachers, and students are the ones who experience the consequences of making good or bad choices in software selection. And they are the ones who most quickly learn the consequences of poor choices.

Criteria for Evaluating E-Activities

The teacher can apply the same assessment techniques used on other instructional materials when they evaluate courseware. The following list is an example:

1. Does the activity meet the age and attention demands of your students?
2. Does it hold the students' interest?
3. Do the programs, activities, or simulation games develop, supplement, or enhance curricular skills?
4. Does the work require adult supervision or instruction?
5. Children need to actively control what the program does. To what extent does the program allow this?
6. Can the courseware be modified to meet individual learning requirements?
7. Can it be adjusted to the learning styles of the user?
8. Does the program have animated graphics which enliven the lesson?
9. Does it meet instructional objectives, and is it educationally sound?
10. Does the activity or program involve higher level thinking and problem solving?

Student Evaluation Checklist

Name of student evaluator(s): _____

Name of software program: _____

Year of software release: _____

Publisher: _____

Subject: _____

1. How long did the activity take? _____ minutes
2. Did you need to ask for help doing the work? Yes/No
3. What skills do you think the activity tried to teach?
   Please circle the word that best answers the question.
4. Was it fun to use? Yes/No/Somewhat/Not very
5. Were the directions clear? Yes/No/Somewhat/Not very
6. Was it easy to use? Yes/No/Somewhat/Not very
7. Were the graphics (pictures) good? Yes/No/Somewhat/Not very
8. Did the activity get you really involved? Yes/No/Somewhat/Not very
9. Were you able to make choices while using it? Yes/No
10. What mistakes did you make? _____
11. What happened when you made a mistake? _____
12. What was the most interesting part of the activity? _____
    _____
13. What did you like the least about it? _____
14. What's a good tip to give a friend who's getting started with this activity?
    _____

Extended Work:

- Make up a quiz about the particular computer program or online activity and give the questions to other students in the class who have used it.
- Create your own soundtrack for part of the activity.
- Make up a student guide for the activity. Use your own directions and illustrations.
- Write or dramatize an interview with one of the characters in a television program or in a movie.
- Interview other students who have used the activity and write their responses.
- Write a review of the activity and post it online.

Remember, there are many "top ten" or "best of lists" out there for consideration to help you choose the best software to fit your needs.

SERVING PEDAGOGY

The curriculum should drive the technology rather than the other way around. The most effective use of such tools occurs when desired learning outcomes are figured out first. Once standards have been set, you can, then, decide what technological applications will help you reach your goals. Technology can help when it serves clear educational goals. But there is much work to be done; put an "e" and a dash in front of education. However, by working together, teachers can learn from each other, push on to other innovations, and build a medium worthy of our students.

Converging communication technologies can serve as a great public resource. Commercial profit alone should not determine how these new technologies will be exploited. The airwaves and the information highways are owned by the public and should serve the public. Emerging public utilities must show a concern for learning and responsible social action. Any new media system is a public trust that should enable students to become intelligent and informed citizens.

In today's world, children grow up interacting with electronic media as much as they do interacting with print or with people. Unfortunately, much of the programming not only is violent, repetitious, and mindless, but also distracts students from more important literacy and physical exercise activities.

New technologies, freshly engineered software, and interactive media amplify everything. To deal with these new digital realities requires a new approach to curriculum and instruction. It also requires a heightened sense of social responsibility on the part of those in control of programming.

Not only has the media changed, but we now live in a society that is ruled by profits. It is the time to go back to considering the public interest and devise rules to insure that the mass media takes responsibility for the fact that they influence the foundation on which formal learning takes place.

Tips on Using Digital Technology

- Technology is an important thing, but not the only thing.
- Put the learning piece in place first.
- Don't panic or be intimidated by technology.
- Make mistakes and learn from your failures.
- Share your successes with other teachers and students.
- Don't be afraid to learn from your students.
- Bring everybody along; leave no one behind.
- Collaborate with other teachers; invest in teacher training.
- Two-way is the only way. Foster active conversation.
- Have fun. Be sure that there is some joy in it.

Remember, both real and imaginary factors can stir people to action.

The human imagination can be enhanced by technology-based instruction in a way that makes actual experiences more meaningful. With computer control, you can speed it up, slow it down, go into an atom, or go back into the past. On occasion, the experience can even transcend print or actual experience as an analytical tool.

Pushing the Boundaries of Knowledge and Change

To move technological possibilities in a positive direction requires learning from the past, living in the present, and building for the future. It also requires asking questions about life in a wired age. For example: "Do modern social networks make relationships more shallow?" or "Do they make the need for face-to-face offline interactions more important?"

Clearly, today's youth need to learn how to "… navigate identity, intimacy, and imagination in a digital world …" (Gardner & Davis, 2013). But no matter how the promises and pitfalls surrounding computers and the Web work out, the undeniable fact is that the digital future in now irretrievably the present.

The human mind can create beyond either what it intends or what it can foresee. Combinations of electronic media will be with us no matter how far the electronic environment expands. The problem is being sure that it works to human advantage. The informing power of technological tools can help change the schools. It can also make learning come alive and breathe wisdom into instructional activities.

Technological application goes a long way toward explaining the shape of the everyone's world today. Along the path to the future, many people will be used to having technology between them and reality. We shape digital technologies and at the same time, they shape us. Along the way, many strands of possibilities shape the future of people's lives. Clearly, the focus of education must be shifted in a way that helps prepare students for a future that is vastly different from what their teachers have known.

We know that technologies like the Internet are fragmenting the thought processes of young people. Also, there is less-and-less time for reflection and quiet thought. Still, it is possible that creative minds may sometimes be strengthened by the ability to interact with the Web and Wikipedia (Thompson, 2013). But no matter where all the changes and apps take us, reasoning, communication, and teamwork skills are bound to go hand in hand with taking an active part in shaping the future.

Technology and engineering have provided us with the tools to live in a digital, global, and fast-changing world. It's a place where educators have to explore and invent new ways to engage students and reimagine their approach to instruction. There are times when the integration of digital innovations into classroom routines can improve our teaching.

Without broad support from society, the schools will not be able provide a good education for all students. New technologies and engineering designs are filled with possibility. Take the example of social media; just because we come up with more and more ingenious ways of communicating with others doesn't insure that we understand what is important or what's being communicated.

Is access to an infinite amount of information infinitely valuable or does it simply lead to the collapse of clear and well-thought-out meaning? Whatever your answer, new media has to have a human face, or uncontrolled technological change will get in the way of educational progress.

Although the technology and engineering path to the educational future may be bumpy, it doesn't have to be gloomy. Clearly, dedicated teachers who understand and know how to use their high-tech tools can make a real difference.

We may live in an information age, but a clear understanding of the nature of information remains elusive. With or without an understanding of the concept, increasing the information flow presents at least as many problems as it does opportunities. The same thing might be said about technology in general.

For many around the world, advanced information technology is indistinguishable from magic. But no matter how fascinating the latest digital tools are, it is important to remember that the goal is to develop future citizens who have the skills, values, and tools to understand and question what's going on around them.

What's needed now is a controlling vision that can help explain concepts related to information, technological innovation, and implications for education. But with or without a grand plan, educators can strive to educate the whole person to be productive in a rapidly changing, technology-intensive, and democratic society.

Questions for Readers to Discuss

- How have your digital gadgets changed how you think, live, and work?
- Is it possible to have a free flow of information when you feel that you are being watched and recorded?
- Is social thought control a serious danger in a wired world?
- Can computer algorithms predict personal or criminal intentions?
- Is there an algorithmic way to predict who will click, buy, lie, or die?

- Should the government (NSA) be allowed to piggyback on the online tools that advertisers use to track consumers?
- How can there be meaningful oversight, transparency, and accountability when it comes to the collection of personal data?

USING ENGINEERING AND
TECHNOLOGY TO EXTEND STEM

Looking back on the lesson plans employed, it's possible to extend STEM contents and processes throughout. Competency in science, technology, engineering, and mathematics is the foundation of creativity. Creativity often occurs when someone has mastered two or more fields and uses that knowledge to think afresh about another. To help coach students through their projects, a student research handbook published by the National Science Teachers Association (NSTA) has been a valuable tool.

STEM teaching recognizes that reasoning skills and creative innovation are vital in any lesson. Personalized STEM learning means doing whatever it takes to provide everyone in the classroom with powerful intellectual tools for imagining new approaches and unique opportunities while maximizing students' learning. STEM instruction pays close attention to who we reach, the content, and how we teach it.

At its best, STEM nurtures the creative spirit in all students as they deepen their understanding. By encouraging students' interests, strengths, and individual preferences, everyone is honored (Bybee, 2013).

How students understand and apply what they have learned is shown in everyday situations. In the STEM classroom, science, technology, engineering, and mathematics are related to real-life situations. Effective teachers provide many opportunities for students to explore, identify, interpret, and apply knowledge. Efforts are valued and students are encouraged to assess themselves.

Now, more than ever, the STEM curriculum attends to responsible citizenship. To some degree, the STEM curriculum can reflect human values and emphasize responsibility by helping student realize they are part of larger social systems such as the global community.

Helping students learn how to make wise decisions and solve difficult problems are among the goals in subjects across the curriculum. New approaches to teaching the STEM subjects stress interdisciplinary work across the curriculum. The following are just a few of the emerging STEM fields: biochemistry, biophysics, global warming, artificial intelligence, plant engineering, terrestrial biology, and neurobiology.

We have trod the face of the Moon, touched the innermost pit of the sea, and can link minds across vast distances. But for all that, it's not so much our engineering ability and technological achievements but what we understand and believe that will determine our fate.

—Paraphrasing Tim Flannery

SUMMARY, CONCLUSION, AND LOOKING AHEAD

Amid all the commercial maneuvering, new media technologies have at least the potential for empowering students to take some control over their learning.

Digital technology can extend students' reach for new knowledge and assist them in their search for new ways of learning. It can also make information more visually intriguing and provide for two-way communication with live or artificially intelligent experts. Tutoring, for example, is often done with live, recorded, or computer-composed experts.

Digital interactive storytelling techniques will allow plot lines to evolve in almost infinite directions. Characters can even be programmed to surprise us in ways that have never been programmed or written. Lessons can play to our cognitive strengths. There are at least three ways to think of spelling a word, for example. One is to picture it. This may be thought of as an experiential kinesthetic approach. Another way is to sound a word out, an audio approach. A third approach is visual. You see the word in your mind and spell it.

It is important to recognize the fact that digital textuality is quite different from what's found in print culture. It has a different logic and takes a different approach to the construction of knowledge. Although digital literacy is still in its infancy, it has already shown real promise and serious limitations. The result is movement toward a new, hybrid cultural landscape.

Computer-generated technology has been used, designed, and engineered in order to give the user a sense of being in another reality. Examples include virtual reality devices and iPhone apps. From airline pilots to surgeons, computer simulations have been used to hone skills for years. But new devices are so powerful that when you put the headset on, you might confuse real experiences with the virtual.

As the technology improves and the cost comes down, real possibilities for learning open up. (Google, *The New York Times*, and others have already come up with a cardboard virtual reality headset viewer that is both cheap and realistic.)

Within the multisensory world of virtual reality, people can see, hear, and touch objects. Even smell can be part of the experience. Some of the applications being developed involve what has been called "telepresence," which gives the operator the sensation of putting his/her hands and eyes in a remote location. One could, for example, send a robot to "Mars" or the bottom of the ocean, control the action, see what it sees, and feel what it feels.

Can we predict where all of these whiz-bang technologies and advanced engineering will take us? The best advice for all of us is to explore multiple paths and not be afraid to change direction. In this way, we can push back the horizon of predictability.

The more unsettled things become, the more important it is to have a range of possibilities. Even luck is the residue of design. If the implications of multiple versions of the future are thought through, it is bound to be good preparation for whatever happens.

Learning to separate noise from knowledge is increasingly important in this age of too much information. Making better use of "knowledge machines" (digital technology) means going well beyond electronic workbooks or surfing the Internet to collaboratively explore problems and tinker with disparate ideas. The good news is that it's at least possible for technological tools to provide a vehicle for building on students' natural curiosity and promoting real learning through active engagement and collaborative inquiry (Gardner & Davis, 2013).

When considered collectively, digital technology is the most powerful representational medium ever invented. It should be put to the highest tasks of society. As far as the schools are concerned, this means using technology to motivate students to explore school subjects with passionate curiosity.

Whether it's in or out of the classroom, many school-aged youngsters in developed countries spend much of their time engaged with media. It often becomes their main source of information, knowledge, and experience. They may say that what really counts is real-life encounters, but the reality is that many are more strongly influenced by online activity.

The engineering (design) of digital devices will continue to evolve in ways that extend human capabilities and compensate for the limits of the human mind. Clearly, dealing with emerging digital technologies is one of the many challenges that teachers face today. We often lament the fact that new teachers have not been provided the training to use new technologies available to them.

Unfortunately, enthusiasm is sometimes lacking when it comes to math, science, and technology. Still, to be successful, teachers need to have at least a reasonable level of comfort with any subject that they are responsible for teaching. And at the primary level, that means everything.

Some teachers are still coming through teacher education programs that did not prepare them adequately to integrate digital technology with instructional tasks. In addition, some school districts pay little attention to the in-service needs of their teachers. This is not a question of remediation. Rather, it is the fact that any profession requires its members to keep up to date with new findings and cutting-edge technology.

College classes and outside experts can help with professional development. But schools need to acquire their own in-house capabilities for helping teachers with professional development.

Increasingly, today's teachers and students have to navigate a world that is networked and global. Being able to work with others on online and offline matters. Collaborators can be sitting next to you, or they can be halfway around the world. In the future, understanding and creating with digital tools will be as important as basic reading, writing, and math skills.

The future of education and technology is bound to be invaded by the present. From social networks to search engines, digital tools will continue to redefine literacy and reshape the architecture of learning. A wide range of new technological tools are already helping students learn in different ways, while permitting them to explore a much broader world of knowledge and learning. There is much more to come.

Perhaps, AI can be used to help enhance and develop the human variety. But since we still don't have a clear understanding of human or machine consciousness, it will take a while to really make a big difference. Then again, it may not take too long. After all, we have already become tone-deaf to the difference between a human and an AI-assisted computer voice.

When it comes to change, it always helps if you are prepared enough to see around some of the corners. Also, everyone involved needs to realize failure is an inevitable cul-de-sac on the road to success. Still, the road to success requires learning from what goes wrong and not making the same mistake twice.

The fields of science, technology, engineering, and mathematics (STEM) are a tangle of collaborations, experiments, misunderstandings, and discoveries. Sometimes, things come together in a calculated way. At other times, new findings are accidental and built on past failures. The process is not always built on pure objectivity; social dynamics, culture, and the personality of those seeking answers are part of the process.

Learning to risk mistakes is part of the learning process. It helps if teachers focus their feedback on the learning processes, and not on a student's personal attributes. Remember, we all build our understandings through trial, error, and tenacity (grit). Sometimes, the STEM subjects help us see things with great clarity—at other times, the findings are blurry.

As far as tomorrow is concerned, we can only begin to imagine how new possibilities and approaches will change the future of learning and life. To successfully sail through the crosscurrents of a transitional age requires the development of stronger intellectual habits and a willingness to learn about what's going on in a rapidly changing world.

Suggestions: What are the most important innovations of the last 120 years? Think about life-altering innovations for your grandparents and for you. How do meaningful past events influence who you are and who you may become?

Try having a conversation with your future self. How do these new technologies and engineering designs fit in?

The future is not some place we are going, but one we are creating. The paths are not just found, but made. And the activity of making them changes both the maker and the destination.

—John Schaar

REFERENCES

Brown, J. & Duduid, P. (2000). *The social life of information.* Boston, MA: Harvard Business School Press.

Bybee, R. (2013) *The case for STEM education; Challenges and opportunities.* Arlington, VA: NSTA.

Firestein, S. (2016). *Failure: Why science is so successful.* Oxford, UK: Oxford University Press.

Flannery, T. (2011). *Here on earth: A natural history of the planet.* New York, NY: Atlantic Monthly Press.

Gardner, H. & Davis, K. (2013). *The app generation: How today's youth navigate identity, intimacy, and imagination in a digital world.* New Haven, CT: Yale University Press.

Kaufman, B. & Gregoire, C. (2016). *Wired to create: Unraveling the mysteries of the creative mind.* New York, NY: Perigree/Penguin.

Levy, S. (2011). *In the plex: How Google thinks, works, and shapes our lives.* New York, NY: Simon & Schuster

Lynch, M. (2016). *Internet of us: Knowing and understanding less in the age of Big Data.* New York, NY: Liveright Publishing Corporation.

Mcluhan, M., & Lapham, L.H. (1994). *Understanding media: The extensions of man.* Cambridge, MA: MIT Press.

Merriam-Webster Dictionary (2004). *The Merriam-Webster Dictionary.* Springfield, MA: Merriam Webster Incorporated.

Moyer, R. & Everett, S. (2012). *Everyday engineering: Putting the E in STEM teaching and learning.* Arlington, VA: National Science Teachers Association.

National Research Council (NRC) (2011a). *Successful STEM education.* Washington, DC: The National Academy Press.

National Research Council (NRC) (2011b). *A Framework for K–12 science education: Practices crosscutting concepts and core ideas.* Washington, DC: The National Academy Press.

National Science Teachers Association (NSTA). Partnership for 21st Century Skills 2009. *21st century skills maps*. Arlington, VA: NSTA. Document available using this link: http://www.p21.org/storage/documents/21stcskillsmap_science.pdf

Organisation for Economic Co-operation and Development (OECD) (2013). Paris, France. www.OECD.org

Ripley, A. (2013) *The smartest kids in the world and how they got that that way*. New York, NY: Simon & Schuster.

Robinson, K. (2015). *Creative schools: The grassroots revolution that's transforming education*. New York, NY: Penguin.

Rose, E. (2010). Continuous partial attention: Reconsidering the role of online learning in the age of interruption. *Educational Technology*, 50:4 July/August, 2010.

Schaar, J. (2016). GoodReads website, from link: http://www.goodreads.com/quotes/279924-the-future-is-not-some-place-we-are-going-but

Thompson, C. (2013). *Smarter than you think: How technology is changing our minds for the better*. New York, NY: Penguin Press.

U.S. Department of Energy. 2005. Wind, www.eere.energy.gov/topics/wind.html

Chapter Five

STEM

Past, Present, and Future

The initial mystery that attends any journey is: how did the traveler reach the starting point in the first place.

—Louise Bogan

STEM instruction is an interdisciplinary approach involving using some combination of science, technology, engineering, and mathematics. Lessons can be structured around two, three, or all four subjects. In a world increasingly filled with surprises, it makes sense to use realistic situations and problems. Also, making connections across the curriculum and thinking about future possibilities is part of the STEM equation.

Using STEM-related intellectual tools is considered a good way to deal with many concepts and subjects. When it comes to dealing with an increasingly complex world, all students can profit from inquiry, problem solving, and understanding the issues surrounding technology. The next step is figuring out how the STEM-related processes might work in day-to-day life.

Interacting imaginatively with others and communicating individual values are all part of quality instruction. Pushing the sphere of what's known is the essence of creativity. Early on, it is important for children to realize that STEM concepts have a role to play in everything from current affairs to music, sports, and literature (Sawah & Clark, 2015).

In the classroom, all learners have the ability to go beyond mastering content to use their imaginations. But the better prepared students are, the more likely they are to be willing to go beyond conventional wisdom, take risks, and learn from their failures. To paraphrase Pasteur, *change favors the prepared mind.*

TEACHING THE STEM SUBJECTS

Many teachers of science and mathematics at the elementary and middle school levels have to teach a wide range of subjects. So, it should not come as a surprise to find out that many teachers (especially at the elementary level) have not received specialist training in many of the core subjects.

Elementary school teachers frequently have to teach the full spectrum of content, so it is expecting too much to expect them to be really well grounded across the curriculum. By the middle grades, it is more likely that there will be more specialists. But whatever the teachers' background is—or the grade being taught—it should be possible to move all students in the direction of inquiry, reasoning, and problem solving (Bybee, 2010).

Science, math, engineering, and their technological associates are simply too important to be left to the experts. Only a small percentage of students will become scientists or mathematicians. But all students must know something about these subjects to become competent workers and citizens.

A major goal of STEM education is to help students develop more self-understanding and move in the direction of responsible citizenship. Along the way, new technologies are altering the way knowledge is conveyed.

New media is one thing, but spending a lot of time working alone in front of a computer screen doesn't make sense. A couple of hours a day (at school and at home) may be fine, but six or seven hours are bound to have negative consequences. In fact, ParticipACTION and other guidelines have suggested that 75 percent of young people exceed any reasonable daily guidelines for screen time.

Increasingly, STEM instruction stresses more interactive (face-to-face) science and math lessons. Today's knowledge explosion moves both the workplace and the classroom in the direction of more collaboration. As students conduct investigations with others and use the tools of technology, they learn how to think critically about the differences between evidence and explanations (Etheredge & Rudnitsky, 2003).

The reach of the STEM education extends beyond literacy and the schools. The evolving nature of science, technology, engineering, and mathematics is one of the reasons for so much debate within the scientific community and in the general public. Although there are certain agreed-upon principles, uncertainty and change go with the territory.

CORE IDEAS AND PRACTICES

There are interdisciplinary areas involved in studying each of the STEM subjects. For example, in science we have life science, physical science, earth/space science, engineering, technology, and the application of science. Some core ideas that cut across these fields include: matter and interactions and energy. Students understand the same concepts that are relevant in many fields. These concepts should become familiar as students progress from grades K through 12.

The various frameworks for the STEM subjects emphasize key practices that students should learn: asking questions, defining problems, and analyzing and interpreting data. Other important practices include explaining ideas and designing solutions. These practices need to be linked with the study interdisciplinary core ideas and applied throughout students' education.

Guiding Principles

1. Learning: Ideas should be explored in ways that stimulate curiosity, create enjoyment of math, and develop a depth of understanding. Students should be actively engaged in doing meaningful mathematics, discussing ideas, and applying math in interesting, thought-provoking situations.
2. Tasks should be designed to challenge students. For example, some short- and long-term investigations should connect procedures and skills with conceptual understandings. Tasks should generate active classroom talk, promote conjectures, and lead to understanding of the necessity of math reasoning.

Standards for Practice

1. Make sense of problems and be persistent in trying to solve them.
2. Reason and draw conclusions.
3. Create reasonable arguments.
4. Model with math and science situations.
5. Use appropriate tools.
6. Be precise.
7. Clearly express your reasoning.
8. Interpret results.
9. Report on the conclusions and the reasoning behind them.

TOOLS OF SCIENCE

Although it is usually best for children to construct knowledge for themselves, we should recognize that students frequently have false understandings about math-related concepts. Some of the misconceptions that students have are natural; some are picked up from the media and the home environment. Just have your students draw a picture of a mathematician or a scientist, and you will have some graphic evidence of stereotypes. You might also have them keep track of representations in film or television.

Children have a natural curiosity when it comes to using one or more of the STEM subjects to examine the natural world. In fact, all young people learn more by experiencing things for themselves, building on what they have already learned, and talking with other students about what they are doing.

Observing, classifying, measuring, and collecting are just a few examples of the processes that learners can learn and apply.

Methods of Thinking and Asking Questions

How students make plans, organize their thoughts, analyze data, and solve problems is *doing* science and mathematics. People comfortable with science and math are often comfortable with thinking. *The question* is the cornerstone of all investigation. It guides the learner to a variety of sources revealing previously undetected patterns. These undiscovered openings can become sources of new questions that can deepen and enhance learning and inquiry.

Questions such as "How can birds fly?", "Why is the sky blue?", or "How are rainbows formed?" have been asked by children throughout history. Obviously, some of their answers were wrong. But the important thing is that the children never stopped asking—they saw and wondered, and sought an answer.

There are times when complete answers can't be found. Fortunately, a deep question is sometimes even better than a clear answer.

STEM AND MAKING CONNECTIONS

- **STEM subjects pursue patterns and relationships.**
  Children need to recognize the repetition of various STEM concepts and make connections with ideas they know. These relationships help unify the science and math curriculum as each new concept is interwoven with former ideas. Students quickly see how a new concept is similar or different from others already learned. For example, young students soon learn how the basic facts of addition and subtraction are interrelated ($4 + 2 = 6$ and $6 - 2 = 4$). They use their science observation skills to describe, classify, compare, measure, and solve problems.
- **STEM subjects are tools.**
  Mathematics and technology are good examples of tools that scientists use in their work. It is also used by all of us every day. Students come to

understand why they are learning the basic science and math principles and ideas involved in the school curriculum. Like scientists, they also will use science and mathematics tools to solve problems. They will learn that many careers and occupations are involved with the tools of various STEM subjects.

- **STEM can be fun (a puzzle).**

Anyone who has ever worked on a puzzle or stimulating problem knows what we're talking about when we say science and mathematics are fun. The stimulating quest for an answer prods one on toward finding a solution.

- **STEM can be a form of art.**

Defined by harmony and internal order, the STEM subjects need to be appreciated as art forms where everything is related and interconnected. Art is often thought to be subjective, and by contrast, science and mathematics are often associated with objective memorized facts and skills. Yet the two are closely related to each other.

Students need to be taught how to appreciate the STEM-related beauty all around them—for example, exploring fractal instances of science and math in nature. (A fractal is a wispy tangled curve that seems complicated no matter how closely one examines it. The object contains more, but similar, complexity, the closer one looks.) A head of broccoli is one example. If you tear off a tiny piece of the broccoli and look at how it is similar to the larger head, you will soon notice that they are the same. Each piece of broccoli could be considered an individual fractal or a whole. The piece of broccoli fits the definition of a fractal appearing complicated; one can see consistent repetitive artistic patterns.

- **The Language of STEM, a Means of Communicating.**

The STEM subjects require being able to use special terms and symbols to represent information. This unique language enhances our ability to communicate across the disciplines of technology, statistics, and other subjects. For example, a young child encountering ($3 + 2 = 5$) needs to have the language translated to terms he or she can understand. Language is a window into students' thinking and understanding.

Our job as teachers is to make sure students have carefully defined terms and meaningful symbols. Statisticians may use mathematical symbols that seem foreign to some of us, but after taking a statistics class, we, too, can decipher the mathematical language. It's no different for children.

Symbolism, along with visual aids such as charts and graphs, is an effective way of expressing science and math ideas to others. Students learn not only to interpret the language of math and science, but also to *use* that knowledge.

- **STEM subjects are interdisciplinary.**
  Students work with the big ideas that connect subjects. Science and mathematics relate to many subjects. Science and technology are the obvious choices. Literature, music, art, social studies, physical education, and just about everything else make use of science and mathematics in some way. If you want to understand what you are reading in the newspaper, for example, you need to be able to read the charts and the graphs taught in science and math classes.

Activities for Understanding

1. *STEM subjects used as methods of thinking.*
   List all the situations outside of school in which you used several STEM subjects during the past week.
2. *STEM subjects used to find patterns and relationships.*
   Show all the ways fifteen objects can be sorted into four piles so that each pile has a different number of objects in it.
3. *STEM subjects used as tools for problem solving.*
   Solve these problems using STEM tools:
   - Will an orange sink or float in water?
   - What happens when the orange is peeled? Have groups do the experiment and explain their reasoning.
4. *STEM subjects solve puzzles and can be fun.*
   With a partner, play a game of cribbage (a card game in which the object is to form combinations for points). Dominoes is another challenging game to play in groups.
5. *STEM subjects can also involve art.*
   Have a small group of students design, paint, or draw different colored geometric figures or fractal-based pictures.
6. *STEM subjects require the use of its own language.*
   Divide the class into small groups of four or five students. Have the group brainstorm about what they would like to find out from other class members (favorite hobbies, TV programs, kinds of pets, and so forth). Once a topic is agreed on, have them organize and take a survey of all class members. When the data has been gathered and compiled, have groups make a clear, descriptive graph that can be posted in the classroom.
7. *STEM subjects lend themselves to designing interdisciplinary activities.*
   With a group, design a song using rhythmic format that can be sung, chanted, or rapped. The lyrics can be written and musical notation, added.

STEM Instruction

In teaching the STEM subjects, it is important to emphasize the processes of good practice, problem solving, reasoning, proof, communications, making connections, and forming scientific and mathematical relationships. Along the way, it is important to motivate students for lifelong learning, while awakening curiosity and encouraging creativity.

In the last century, teachers emphasized the memorization of facts and answering questions correctly. Now, more attention is paid to helping students learn how science and mathematics relate to social problems, technology, creative innovation, and their personal lives.

A new pattern for teaching STEM is emerging that attends to content and the characteristics of effective instruction. Engaging students in active and interactive learning deepens their involvement in their academic work and their understanding of the subjects we teach. Good teachers know that students should have many opportunities to interpret science and math ideas and construct understandings for themselves.

In the latest approaches to STEM, teaching methods are designed make an effort to be purposeful, providing meaningful activities with real applications that touch people's lives. By emphasizing collaborative inquiry, promoting curiosity, and valuing students' ideas, these STEM subjects— science, technology, engineering, and math—become more accessible and interesting.

A COLLABORATIVE MODEL

Views of learning emphasize thinking processes within the learner and point toward changes that need to be made in the way that educators have traditionally thought about teaching, learning, and organizing the school classroom. Central to creating such a learning environment is the desire to help individuals acquire or construct knowledge.

That knowledge is to be shared or developed—rather than held by the authority. It holds teachers to a higher standard, for now, they must have both subject matter knowledge and pedagogical knowledge based on an understanding of learning and child development.

The collaborative learning model for STEM inquiry emphasizes the intrinsic benefits of learning rather than external rewards for academic performance. Lessons are introduced with statements concerning reasons for engaging in the learning task. Students are encouraged to assume responsibility for learning and evaluating their own work and the work of others.

Interaction may include discussing the validity of explanations, searching for more information, the testing of various explanations, or considering the pros and cons of specific decisions.

The characteristics that distinguish today's collaborative learning revolve around group goals and the accompanying benefits of active group work. Instead of being told they need information, it is best if students learn to recognize when additional data are needed.

As students collaborate, they seek it out information, think about it, and suggest possible solutions. Along the way, they come to appropriate the power and value of STEM. In this way, the actual use of information becomes the starting point rather than merely an "add on." The teacher can facilitate the process instead of acting as the sole knowledge dispenser. Student success is measured by performance, work samples, projects, and applications.

Cooperative STEM learning models emphasize exploring a problem, thinking, and the collaborative challenge of posing a solution. This involves peers helping each other, self-evaluation, and group support for risk-taking. This also means accepting individual differences and having positive expectations for everyone in the group.

As learners can come to understand the purpose of their tasks and contribute to their own learning and the learning of others, the end result is greater group persistence and more self-directed learning.

Emphasizing group-based STEM instruction allows students to address common topics at many of levels of sophistication. The usual procedure is to have all students work on the same topic during a given unit. The work is divided into a number of investigatory or practical activities tackled by student teams.

Activities for STEM instruction can include basic required work and optional enrichment work, so that less able groups should accomplish the requirements and be able to choose from some of the options. The more able groups should move on to more challenging assignments after completing the basic tasks. Because topics are not sequenced linearly, each new topic may be assessed as the lesson progresses (Flanagan & Altonso, 2016).

THE INTERNET AND STEM INSTRUCTION

Tim Berners-Lee and others helped make it possible for everyone to use the Internet with the World Wide Web.

In his 2016 film *Lo and Behold*, Werner Herzog suggests that opening the Internet up for all is as momentous as the introduction of electricity into our civilization. Is it doom, gloom, "glorious," or all of the above?

Although some of the elements had been around for years, nobody really saw today's Internet coming. If you carefully looked around in the 1980s, you might have picked up some hints. But even as late as the early 1990s, everyone from Microsoft to science fiction writers missed the implications. Tomorrow's STEM-influenced realities will prove just as difficult to predict.

To paraphrase Andrew Russell (2016), there are at least *five myths about the World Wide Web*:

1. We know who invented the Web and the Internet, and when.
2. The Web is an American invention.
3. Government power is obsolete on the Internet.
4. The gatekeepers are dead and everything is disrupted.
5. A massive cyberattack is coming.

A good activity is to fill in the detail on each point and explain where you agree and disagree.

> *Thanks to social media and an increasing flood of data, the capacity to generate causes and controversies almost instantly has become the norm in today's "super-transparent" society.*

—Austin & Upton

ACTIVE TEACHING AND LEARNING IN A POST-TRUTH WORLD

There is general agreement that a constructive, active view of learning must be reflected in the way that the STEM subjects are taught. Classroom experiences should stimulate students, build on past understandings, and encourage students to explore their own ideas. This means that students have many chances to interpret concepts and construct understandings for themselves.

The human brain has the ability to scan a vast warehouse of information, knowledge, and past experiences to come up with unique solutions to problems. For many problems on the horizon, it is important to remember that coming up with solutions often takes longer than you thought it would—and then, change happens faster than you thought it could.

To paraphrase William Davies, facts are a type of knowledge can reliably be given out without the constant need for verification or interpretation. We live in a world of too much data and too few agreed-upon "facts." Today's reality seems to be more defined by situations and opinions than numbers or statistics. Caught between social media and populist movements, we may be entering a postfactual (or posttruth) age.

Thinking and reasoning in today's STEM subjects have a lot to do with how things are going to develop in the future. Energizing students in a way that helps them grow into informed participants in tomorrow's social and technological changes requires engaging them in real-world problem-solving investigations and engaging projects.

A good approach to teaching the STEM subjects involves working with a partner or a small group. The basic idea is to get feedback on problem-solving work. To encourage peer review and joint authorship, we suggest that students keep a daily log or journal. As students talk together, they can better understand what they have been working on.

With the renewed emphasis on thinking, communicating, and making connections among topics, students are more in control of their learning. With collaborative inquiry, students have many experiences with manipulatives, calculators, computers, and working on real-world applications.

There are more opportunities to make connections and work with peers on interesting problems. The ability to express basic math and science understandings, estimate confidently, and check the reasonableness of their estimates is part of what it means to be literate, numerate, and employable. For example:

- Telling stories is enhanced by having children use unifix cubes or other manipulatives to represent the people, objects, or animals in the oral problems.
- Have children work on construction paper or prepare counting boards on which trees, oceans, trails, houses, space stations, and other things have been drawn.

Making Interdisciplinary Connections

Mathematics and technology have always served as important tools for work in the sciences.

The STEM subjects inform everything from history to the evening news. They enrich the visual and performing arts as well as sports and physical education. As an extension of our natural language, they provide a context for language learning.

As STEM-related concerns become more integrated into society, their interconnectedness with other school subjects becomes an important goal.

INTERDISCIPLINARY STEM ACTIVITIES

We'd like to get away from the idea of splitting up the curriculum; instead, the new focus is looking at fusing various disciplines together. The next few activities try to accomplish this goal.

Activity 1: Using the Sun to Teach Geometry

Objectives:

Students will:

- Describe, model, draw, and classify shapes.
- Investigate and predict the results of combining, subdividing, and changing shapes.
- Develop spatial sense.
- Relate geometric ideas to number and measurement ideas.
- Recognize and appreciate geometry in their world.

Directions:

1. Have students investigate figures and their properties through shadow geometry, exploring what happens to shapes held in front of a point of light. They can also explore what happens to shapes held in the sunlight when the sun's rays are nearly parallel.
2. Have children discover which characteristics of the shapes are maintained under varying conditions. For this activity, provide pairs of children with square objects such as wooden or plastic squares.
3. Take students to an area of the playground that has a flat surface. Have the children hold the square regions so that shadows are cast on the ground. Encourage the children to move the square regions so that the shadow changes.
4. Have students talk about the shadows they found. Discuss how they were able to make the shapes larger and smaller. See what other observations they have made.
5. To make a permanent record of shapes, have a student draw a shape on a piece of paper and put the piece of paper on the ground. Let the shadow fall on the paper. Have that student draw around the outline of the shadow.
6. When each student has had a chance to draw a favorite shape, there will be a collection of interesting drawings that can serve as a source for discussion, sorting, and display.

7. For a challenge activity: students may also wish to discover if they can make a triangle or a pentagon shadow using the square region. Using outlines, children drew of the shadows cast by square shapes see if students can find things that are alike and different in the drawings.
8. Have students count the number of corners and sides of each shape and compare those numbers. Encourage them to discuss and record their conclusions.

Precaution: Be sure you direct students not to look at the sun directly.

Activity 2: Design an Escher-Type Art Drawing

Students become aware of the properties of shapes through many experiences. They manipulate, visualize, draw, construct, and represent shapes in a variety of interesting creative ways.

Objectives:

Students will:

• Create patterns.
• Explain their designs.

Directions:

1. Introduce the idea of making a repeated pattern (tessellation). Teachers may wish to talk about artists who use the concept of tessellation in their art work.
2. Have students represent a three-dimensional object on paper.
3. Ask students what shapes can be seen in different objects.
4. Have them try to make a symmetrical design. What shapes will tile a floor (or tessellation)?
5. Begin by having students make a tessellated design. Encourage students to explain their pattern and the relationships between the figures they've chosen. Students who are adept at tessellating (drawing repeated patterns) may wish to extend that skill in artwork designs.

   Through direct experiences with three dimensional objects and then transferring those objects into a two-dimensional world, students become aware of the relationships and properties of geometric shapes. They are able to notice symmetry in the designs they create. They investigate how patterns look if they're moved or rotated. They draw, build, and describe many shapes from a variety of perspectives.

Activity 3: How Long Are You?

Objectives:

Students will:

* Collect, organize, and describe data.
* Measure and compare you and your partner's body parts.
* Explain your measurements.

1. With a partner measure and compare the following body parts:

	You		**Partner**	
	Inches	Feet	Inches	Feet
a. Head to toe:				
b. Arm span: fingertip to fingertip:				
c. Forearm:				
d. Foot:				
e. Circumference around head:				
f. Hand (pinky to thumb):				

2. Explain your measuring technique and state your measurements.

3. Artwork: Draw yourself and your partner's measurements and color.

4. Comparison sentences: What did you learn?

_____

_____

5. Share with the rest of the class.

TEAM EXPERIENCES HELP ATTAIN GROUP KNOWLEDGE

The ability to work in small groups is an important skill. Being able to function as part of a team is something that is important at any age. This is as true in math and science as anywhere else. In the world outside of school, projects that rely on science and math often build on different points of view within the group to produce a coherent whole (Eichinger, 2004). In and out of school, groups functioning as a team are also more likely to persist in working out a problem. As a consequence, they are more likely to succeed.

Most tasks can be made more interesting by taking a team approach. Small cooperative groups can often handle more sophisticated problems better than individuals working alone. Arranging students into teams allows teachers to channel energy into productive communication and problem solving. Such an arrangement gives students an opportunity to take more responsibility for their own learning. Of course, there is no freedom in a vacuum.

Even when students are working in small groups, the teacher guides, mentors, and advises them. It is the teacher who sets up groups and encourages everyone to take responsibility for themselves and for other team members.

When students focus on their own investigations, discussions, and group projects, the teacher's role shifts to that of an expert manager. The whole class setting can be used for initial brainstorming, giving directions, summarizing data, reviewing different strategies, and coming to common understandings about the questions that come up.

A good teacher knows how to ask the right question in different situations. Some of these questions will be discussed in small groups and some with the entire class. Teachers may also want to use the larger class setting to summarize and highlight important ideas that come from the work of the small groups.

When small collaborative groups are combined with an appreciation of the power and beauty of science and mathematics, the fear surrounding these subjects begins to recede and student understanding occurs.

Call it collaborative, cooperative, or team learning, the essential point is that strategies for teaching science and mathematics that involve social interaction result in children gaining more control of the math and science curriculum. Students are willing to ask hard questions and take risks in a supportive caring community.

CHALLENGING STEM ACTIVITIES

This section brings together the science and math standards to elementary and middle school classrooms. Meaningful activities that employ the core ideas and practices of observing, comparing, measuring, recording data, and making good conclusions are emphasized. These activities are based on the 2010 National Framework for K–12 Science Education Standards. Whenever possible, the Mathematics Standards (2011) are included.

Activity 1: Mysterious Stories

Stories are a means of communication. Even before written language, people drew pictures on walls of their caves to show a successful hunt or the animals they met. Stories told of rules and accomplishments. Whether found in dance, pictures, art, or words, it is important to view stories as a desire for people to communicate their thoughts, dreams, and mysteries across generations.

Looking at stories as mysteries is exciting, providing students with characters they can identify who allow them to be included and be part of the adventure.

Activity 2: Invent a Mystery Story

Create your mystery story. Choose a topic of interest to you.

Activity 3: The Mystery of Gravity and Yo-Yos

Content standards:

- Applying the core ideas of physical science.
- Comparing movement and gravity.

Objectives/importance:

Students observe objects concerning force, acceleration, friction, and gravity. Students apply physics to real-world situations.

Background information:

Students should review the basic physics principles of gravity and inertia.

- Gravity: a natural force of attraction that tends to draw objects together
- Inertia: a property of matter whereby it remains at rest or continues in uniform motion unless attracted by some outside force
- Velocity: rate of change of the object's position

Questions:

- How do yo-yos reflect physics?
- What is an example of inertia?
- How does your yo-yo show force?
- Can you make your yo-yo accelerate?
- What causes friction when using a yo-yo?
- Can you explain why a yo-yo shows gravity?

Other student activities:

Students work in pairs with their yo-yos.

1. Estimate the direction your yo-yo moved in one minute.
2. Record the velocity of your yo-yos. What speed do you think they traveled?
3. Write a mystery story of your yo-yos and how they demonstrate gravity.
4. Why does a yo-yo act as it does?
5. There are some professional yo-yo groups. Find out more about them. Record your findings.

Accelerated student activities:

1. Slow down your yo-yo, then try to speed it up.
2. What happens when the yo-yo is thrown in a different direction?
3. What makes the yo-yo slow down?
4. When does the yo-yo move fast?
5. Describe the physics involved in these activities.
6. Record your yo-yo movements. Compare your experiences with those of other students.
7. With other students, describe the force, speed, gravity, and friction of your yo-yos. Record your findings.
8. Make a class chart so other classes can see your physical science achievements.

Activity 4: Flowing Mysteries

Grade level: elementary, junior high school.

Content standards:

- Applying the physical science standards.
- Using basic ideas and procedures in science.
- Putting personal views to practical use.
- Practicing oral and written communication skills.
- Employing science and technology skills.

Procedures: Students are experimenting with household chemicals.

Materials:

Student tools for each workstation:

6 small plastic bottles
6 flowing substances (vinegar, soap, alcohol, cooking oil, vanilla, water)
6 short glass tubes with rubber bulb
1 wide-mouthed drinking cup
1 small drinking cup
1 flat-bottomed container for holding articles
1 piece of saran wrap
1 sheet of aluminum foil
1 sheet of waxed paper
1 sheet of white paper

Directions:

1. Prepare the containers with food coloring added.
2. Have students discuss how scientists perform experiments.

3. Prepare the trays and tools for each group.
4. On the chalkboard list the experiments that students might try:
   • Substance races
   • Floating ability
   • Density
   • Combining liquids
   • Other suggestions

Student Task

Discover which of the six chemicals are using the following rules:

• Use your sense of sight to find out what the flowing mysteries are.
• You are not to smell, touch, or taste the chemicals.
• Each dropper can be used to pick up only one substance,

Students rotate among the workstations, experimenting as they try to discover what the flowing mysteries are. They conduct several tests during the process.

Workstation 1 Substance Race

Lesson steps:

1. Choose a substance and a sheet of paper to cover your tray (waxed, white, aluminum foil, Saran wrap).
2. Place a drop of each substance on the paper.
3. Tip the tray so that the substance moves.
4. Record the movement of the chemicals.
5. Try the experiment with all the flowing chemicals.

Workstation 2 Floating Ability

Lesson steps:

1. Select a small plastic container.
2. Add drops of each colored substance.
3. We want to see which substances will float.
4. Encourage students to experiment.
5. Jerk, shake, and maneuver the container to detect which chemicals move to the top.
6. Record the movement of the chemicals.

Workstation 3 Density

Lesson steps:

1. Select a small plastic container
2. Add drops of each colored substance.
3. We want to see which substances will sink.
4. Encourage students to experiment.
5. Jerk, shake, and maneuver the container to detect which chemicals sink.
6. Record the movement of the chemicals.

Workstation 4 Mixing Chemicals

Lesson steps:

1. Guess what each substance is and test your guesses by observing which substances blend together.
2. Record your guesses:

   Blue chemical _____

   Green chemical _____

   Red chemical _____

   Yellow chemical _____

   Purple chemical _____

   Clear chemical _____

   Which will mix?   Write your reasoning:

   Oil: _____

   Soap: _____

   Water: _____

   Vanilla: _____

   Vinegar: _____

   Alcohol: _____

People work in groups, talk, listen, and express ideas. Students share information, explain ideas, and help each other.

Moving the Furniture

When it comes to science experiments and arithmetic computations, students must learn how to work together to put the skills they learn into practice—now and in the future.

Student learning teams are a powerful way to approach mathematics and science instruction. To help students achieve a deeper understanding, more attention is being given to application and social interaction. Collaborative inquiry and problem-solving activities are important routes to deeper understandings of science and mathematics (Jadrich & Bruxuoort, 2011).

In tomorrow's classrooms, interactive learning by small groups of students will be the norm. This makes sense because peer support helps learners feel more confident and willing to make mistakes and engage in scientific inquiry and problem solving in mathematics. Also, collaboration has a lot to do with how scientists and mathematicians work today.

When it comes to science experiments and arithmetic computations, students must learn how to work together to put the skills they learn into practice now and in the future. Technology, teamwork, and idea creation are all part of today's science/math package. So, in the classroom, get used to the furniture and the ideas getting moved around. You may question the educational significance of the Internet, but science, math, and technology lessons are more and more likely to be integrated across the basic curriculum.

UPDATING BELIEFS WITH NEW DATA

No one can make really accurate predictions about the future. Still, Tetlock and Mellers suggest that some people can think more clearly about the direction of change (Mlodinow, 2015). They suggest that the best strategic planners have above average levels of "fluid intelligence" on tests that measured critical and abstract thinking.

Among helpful skills and attitudes:

- An interest in intellectual challenges.
- A clear grasp of geopolitical concepts.
- The ability to collaborate, while avoiding "group think."

A willingness to update beliefs based on new information was cited as the strongest ingredient of accurate predictions (Tetlock & Gardner, 2015).

The insights of modern science have come about with the help of mathematics and technology. Scientific thinking involves testing ideas through experimentation and a creative search for solutions. With the help of mathematical and technological tools, the processes of scientific reasoning are bound to lead us to new discoveries and a better understanding of the world.

As far as mathematics instruction is concerned, teachers have to go beyond teaching students basic arithmetic, how to balance a checkbook, or estimate how long it will take to get from one town to another. Students should not avoid asking big questions, nor should they be afraid to consider using the thinking skills found in the STEM subjects to explore broad trends and related human problems.

As far as specific classroom content is concerned, it is not enough to teach skills and specific facts in isolation from the situations that require those skills. Those who teach the STEM subjects also have a responsibility to meet the challenges of technology and society, while paying attention all the subjects in the school curriculum.

SUMMARY, CONCLUSION, AND LOOKING AHEAD

To thrive creatively in the classroom, it helps if both students and teachers can see the unique work of others. It also helps administrators value classroom environments where imaginative work is the norm. Capacity and subject matter knowledge matter, but creative individuals tend to have a risk-taking personality and temperament. Still, like the idea of multiple intelligences, there are many kinds of creativity, and there are multiple paths to imaginative ideas (Sinclair, 2006).

Whatever approach you take, since failure is widely recognized as part of the creative process, it is especially important to know how to learn from mistakes (Traig, 2015). Although every idea generated is not going to be a good idea, the more new ideas are generated, the more likely it is that something unique and useful will turn up.

In today's classrooms, it is important that students know how to use a wide range of STEM-related tools to learn and apply what's been learned in an imaginative way. It is also important to have some idea how subject or combination of subjects impacts life on a daily basis. At any grade level, competency implies having some idea about how to use knowledge in purposeful ways.

In a world increasingly filled with the technological products of scientific inquiry, engineering and mathematical problem solving makes understanding the implications more important than ever. Clearly, being naive or afraid of any of the STEM subjects can be a real problem in school, in the workplace, and for citizens in a democracy.

Making specific predictions about the future is for fortune tellers and astrologers. Considering the possibilities and identifying themes that could be developed make more sense. It is also important to consider the fact that the speed of change can be just as important as the direction. Ambiguity and the unexpected may rule, but you can sure that new realities will be filled with both promise and anger.

A good way to understand what is just beyond the horizon is to look around today. Remember, history rhymes more often than it repeats itself. Trying to predict the future based on past experiences has its limits. We have to realize that what we see today is the new and the old mashed together, and the future may contain the old and the new. Getting ready for an uncertain tomorrow includes realizing that it's not possible to anticipate every challenge.

The future is already here. It's just not evenly distributed yet.
—William Gibson

REFERENCES

Austin, R. & Upton, D. (2016). Leading in the age of super-transparency. *MIT Sloan Management Review* 57:2, 25–32.

Bogan, L. (1980). *Journey around my room: The autobiography of Louise Bogan.* New York, NY: Viking Press.

Bybee, R. (2010). *The teaching of science in the 21st century.* Arlington, VA: NSTA Press (National Science Teachers Association).

Eichinger, J. (2004). *40 Strategies for integrating science and mathematics instruction: K–8.* Upper Saddle River, NJ: Prentice Hall.

Etheredge, S. & Rudnitsky. A. (2003). *Introducing students to scientific inquiry: How do we know what we know?* Boston, MA: Allyn & Bacon.

Flanagan, D. & Altonso, C. (2016). *Essentials of WISC-V assessment.* San Francisco, CA: Jossey-Bass (Wiley).

Gibson, W. (2012). *Distrust that particular flavor.* New York, NY: G. P. Putnam's & Sons.

Jadrich, J. & Bruxuoort, C. (2011). *Learning and teaching scientific inquiry: Research and applications.* Arlington, VA: NSTA Press.

Mlodinow, L. (2015, October 15). "Mindware" and "Superforecasting." *New York Times.* From link: http://www.nytimes.com/2015/10/18/books/review/mindware-and-superforecasting.html

Sawah, R. & Clark, A. (2015). *The everything STEM handbook.* Cincinnati, OH: F+W Publications.

Sinclair, N. (2006). *Mathematics and beauty: Aesthetic approaches to teaching children.* New York, NY: Teachers College Press.

Tetlock, P. & Gardner, D. (2015). *Superforecasting: The art and science of prediction.* New York, NY: Crown.

Traig, J. (Ed.) (2015). *STEM to story: Enthralling and effective lesson plans for grades 5–8.* San Francisco, CA: Jossey-Bass (Wiley).

Chapter Six

Evaluating STEM Learning

Informative Assessment, Lesson Plans, and Activities

Don't let the instructionally perfect prevent you from reaping the rewards of the instructionally possible.

—Popham

This chapter is designed to help educational practitioners who want to improve STEM lessons before, *during*, and after instruction. We include suggestions for differentiated lesson plans and *in*formative assessment techniques. It is our belief that a well-developed lesson requires a written framework to guide instruction. A good lesson plan can also provide the teacher with a reservoir of well-reasoned questions, activities, paths for exploration, and alternative assessment techniques.

When it comes to assessment, the focus here is on improving learning rather than giving tests. Most tests do not address divergent thinking, creativity, or collaborative problem solving in realistic situations. If a school district insists on teaching to norm-referenced tests, teachers will have trouble designing lessons that reflect a rich and deep science or math curriculum.

As far as teacher-made tests are concerned, there was a time when the main approach was to give an exam at the end of a lesson or unit and make a decision about where to go next. The results have been found to be better when teachers assess along the way so that they can track performance, spot trends, and tailor instruction (as it is happening) to meet the needs of students (Scherer, 2016).

Both students and teachers can profit when lesson plans provide room for making evidence-based decisions about learning as it is actually happening. Popham and others call it a "*trans*formative" framework for assessment and go on to suggest that it can help teachers teach better and students learn better

(Popham, 2008). But no matter what your point of view is, it makes sense for teachers to plan what they are going to do in a way that allows for ongoing adjustments in instruction, classroom climate, and each student's framework for learning.

Whether assessment is before, *during*, or after a lesson, it helps if teachers have time to share ideas about what they are planning and how related assessments fit in. Individuals respond favorably when colleagues take their ideas seriously and constructive reflections on their work are fed back to them. As it is now, some teachers spend almost as much time planning alone as they do teaching and talking with other teachers.

Finding a way to share ideas and lesson plans can improve the quality and actually cut the time involved in planning. With a little help from their friends, teachers can more rapidly prepare lesson plans in a way that helps keep the focus on the objective or destination. It can also help them do a better job of dealing with the different reasoning and problem-solving routes taken by students.

DEVELOPING GUIDES FOR STEM INSTRUCTION

Lesson plans are patterns and formats for supporting lessons. Some schools use curriculum guides and textbooks as organizing elements for science and math instruction. Both can serve as a starting points or guiding principles for lesson planning. New teachers often turn to the *teachers' edition* of the textbook for help with planning; although creative adjustments have to be made, suggested lesson plans are often there for immediate use. If your school district doesn't have what you need, publishers, libraries, and the Internet can help.

There is no consensus about the "best" format for a lesson plan. Ask five teachers and you will probably get five very different answers. Go to the Internet and you will get many more. For practice and lesson preparation, pick two plans that lend themselves to differentiation and do an overlapping Venn diagram (two overlapping circles); place the similarities in the center of the overlap and place the differences on the outside. Share the results with other teachers. There is no good reason to stick with one format. However, you should note that the majority of lesson plans contain the following points: objectives, skills emphasized, materials, procedures, and assessment. The questions teachers like answered:

1. "What are the learners supposed to learn?"
2. "How will students learn it?"
3. "How will I know if they have successfully completed it?"

The sample lesson plans and assessment strategies presented here follow a differentiated instructional model that has been field tested with student teachers and in-service teachers. They are research-based and standards-driven. The lessons are arranged so they can be altered to fit in with any K–8 classroom program or science/math curriculum. The main goal is to help teachers plan and implement long-, medium-, and short-term STEM objectives. The teacher's edition to the textbook may help, but it is no substitute for creative planning on the part of the teachers. In planning lessons, the classroom environment, the materials available, the diversity of the class, the curriculum, and assessment strategies all come into play.

It always helps if students have plenty of practice in applying new skills and knowledge in unique ways. Such a differentiated approach for activities, lesson planning, and informative assessment can help bring focus to a quality science or math concept, lesson, or unit.

A STEM lesson plan might include:

- *Objectives:* the purpose or lesson goal stating what the student should be able to do after the lesson. Consider multiple intelligences and learning profiles: interest, readiness, and multiculturalism.
- *Content standards:* What students should know and be able to do: unifying concepts in science and math, inquiry, life, earth and space, and physical science.
- *Procedures:* the instructional activities and ways of getting students involved in learning the skill being taught. Activate and engage, explain, and apply learning.
- *Materials:* what the student, the learning group, and the teacher need in terms of tools to successfully complete the lesson.
- *Assessment:* how teachers decide if the students have achieved what they wanted them to learn. Informative assessment is an ongoing process. Use rubrics for self-evaluation; portfolio entries should show work and progress over time.
- *Accommodations:* ways of adapting content, materials, and assessment methods so that all students are provided with many ways of learning.
- *Evaluation:* the rating of students' performance, group collaboration, the success of the lessons, and teacher effectiveness.

Getting students comfortable with differentiated science/math ideas and related instructional activities is a road to higher levels of education and a thoughtful life. Teachers often start a lesson by posing a thought provoking question or choosing a problem for a collaborative group to work on. Sometimes, the teacher puts forward some possibilities and the pair or small group of students decides which one to explore. At times, the students come up with the questions or problems that they would like to explore. Whatever the approach, it is important to leave room for ideas and questions, rather than answers. This way, students will be encouraged to activate their own thinking and consider the thinking of others.

QUALIFIED TEACHERS AND QUALITIES OF THE MIND

Schools alone may not be able to close achievement gaps rooted in broader economic and social inequalities, but they can make a big difference. At the classroom level, good teachers can make all the difference in the world when they are given the educational resources needed to thrive: decent facilities, smaller classes, and a rich science/math curriculum. When it comes to instruction, the most important resource is having a *fully qualified teacher.*

In today's typical elementary and middle school classroom, the range of academic diversity makes it very difficult for teachers to teach and assess using the same unaltered procedure for everyone. The traditional direct instructional lesson/testing model doesn't not work for many students. For advanced students, it can lead to boredom, especially if they have already mastered the content.

When capable students are bored, they tend to get off task or even become disruptive. For struggling students, the traditional model often fails because the skills needed to do the lesson are missing. As a result, the material is frequently misunderstood and there is no engagement with the lesson. All too often, the end result is disinterested and disorderly behavior (Chapman & King, 2008).

Providing all students with a safe, structured classroom setting sets the table for learning both appropriate behavior and content. Whatever their academic standing, if you give learners a good reason for doing something, they will probably put up with a little drudgery. But whether students are doing well in science and math or not, it helps if time and space are set aside for exploration and self-discovery. Whether it involves observation, evidence, formulas, comparisons, cause–effect relationships, conclusions, or anything else, increased collaboration and participation can deepen students' understanding of academic content. As cognitive science and neuroscience suggest, building on the social nature of learning and exploring the how's and why's of a subject make it possible for students to unravel problems like a puzzle, rather than memorizing facts or processes for no apparent reason (Sousa, 2008).

In terms of practical advice, we used to think that neuroscience had little to offer and that educators should pay closer attention to cognitive science, psychology, and associated concepts like multiple intelligence (MI) theory.

Others argued that both cognitive psychology and MI theory both have a lot to do with brain science. Of course, the brain and the mind affect everything we do at school, so they have always been natural topics of interest. It's just that now brain research has reached a stage where it has some practical implications for instructional planning. At the very least, neuroscience research can give teachers some ideas about what not to do. Look for more in the future.

Howard Gardner builds on research in cognitive science and psychology to suggest five uses of the mind that will matter even more in the future:

- The *disciplinary mind*—mastery of major schools of thought, including science and mathematics.
- The *synthesizing mind*—ability to integrate ideas from different subjects.
- The *creating mind*—capacity to deal with new problems and questions.
- The *respectful mind*—an appreciation of differences among human beings.
- The *ethical mind*—taking on the responsibilities of citizenship and productive work.

Gardner uses this framework to suggest that the world of the future will demand capacities that, until now, have been options. He also points out that standard measures of learning do not adequately address these "five minds of the future." In addition, norm-referenced tests do not do a good job of measuring whether or not the student has actually learned something that they can put into use outside the classroom door (Gardner, 2006).

TIERED ACTIVITIES AND
DIFFERENTIATED APPLICATIONS

Tiered assignments and assessment are different learning tasks and assessments that teachers develop to meet students' needs. When tiered tasks are used with flexible groups, it usually results in a better instructional match between science/math lessons and individual student needs. In addition, student understanding that comes from the use of differentiated activities are also more likely to transfer to other situations (Tomlinson et al., 2008).

Tiered activities can be arranged by the challenge level based on levels of thinking. They can also be tiered by complexity. Complex tiered activities are designed to be used by levels of difficulty—least complex, more complex, most complex. Another way to tier activities is by learning abilities. When you tier activities by learning abilities, you are matching students' ability levels to their learning needs. Accelerated learners as well as those who have difficulty need to be reminded that every student can be successful.

Teachers use tiered activities so that all students can focus on essential understandings and skills, but at different levels of complexity, open-endedness and ability. By keeping the focus the same, but provoking different ways of reaching the destination, each student is appropriately challenged. Achieving some mastery in several disciplines certainly helps when it comes to integrating ideas from different fields into a coherent whole. Also, intellectual curiosity and creativity are often generated when two or more fields have been mastered and the framework from one is used to think afresh about the other.

We cannot replace curiosity with mechanical memorization and call it knowledge.

—Paulo Freire

MAKING ASSESSMENT AND PLANNING COUNT

Informative assessment can guide instruction in areas where students are having trouble, much like a meaningful homework assignment does. Examining the reasoning behind errors, as well as correcting approaches, can help teachers and students gain a better understanding of science and math. Whether you call it formative, informative, or transformative assessment, teachers can adapt instruction by *looking at students' ongoing work* and make changes that will immediately benefit students. In addition, learners can use their current work samples to actively organize and adjust their own learning (Stiggins et al., 2006).

A good lesson/assessment plan includes:

1. *Looking at* whether students have understood the concept.
2. *Informally assessing* students with one or two quick questions.
3. *Discovering* which students will need alternative instruction.

The teacher, based on his or her judgment and previous experience with the students, "looks" at which students may have grasped the concept. Then, the teacher quickly "assesses" them informally, with a question such as "Do you understand this idea?" *Discovering* requires a bit more explanation. In a typical STEM class, instructional groups will be formed that for active participatory learning.

Using three or four groups in the classroom means that teachers must assume that students can learn from each other. It's much more than looking for the "right" answers; it's debating, revising, and replacing ideas about scientific findings and mathematical operations. Both teachers and students need to be comfortable with a little trial and error to function well within such a collaborative learning and assessment model.

Informative assessment can take a variety of forms in your classroom. It involves helping students answer these questions:

- What work I'm expected to do?
  Give students a list of learning outcomes they are responsible for. Show them some examples of strong and weak performances they are expected to create and decide which one is better and why.
- What am I good at and what do I need to improve?
  Have students informally assess their strengths and areas needing improvement. Give students a simple quiz to help both the teacher and students understand who needs to work on what. Share the results with students. Underline words or phrases that reflect specific strengths and areas needing improvement. Keep a record of learning outcomes and have students check off the ones they've completed.
- What did I accomplish?
  Give students feedback on their work and have them use it to set personal goals. Encourage students to comment on their progress—what changes have they noticed, what is easy now that used to be difficult? When students use teacher feedback and learn how to assess their work and look at future goals, they become very involved in their own success. Teachers and students collaborate in an ongoing process using assessment information to improve their learning.

In order to demonstrate (in)formative science/math assessment practices, we designed the following differentiated lesson/assessment plan model. It involves objectives, rationale, prerequisite skills, content standards, organization, and procedures. Each lesson includes materials, student involvement, group arrangements, adaptations, methods for differentiating by gearing up (if the lesson is too easy) or gearing down (if the lesson is too hard), and formative assessment strategies.

DIFFERENTIATED LESSON/ASSESSMENT PLAN MODEL

Topic:

Grade level:

Objectives: What do you want students to learn?

Rationale: Why are the important concepts/skills?

Prerequisite skills/understandings: What background information do students need before starting?

Content standards: Have you included the national and state standards for your subject area?

Organization and procedures

List the materials needed:

-
-
-

How are you going to get the students involved?

Introduction/anticipatory set:

Lesson development questions:

Closure: How will you end the lesson:

Small group options:

Adaptations: Gearing up (if the lesson is too easy)

Gearing down (if the lesson is too hard)

Formative assessment: Does your assessment provide evidence of learning the objectives? What evidence will you use to determine that what was taught has been learned? (e.g., observations, products produced, portfolio entries, reflections.)

Creating Effective Informative *Assessment:* To connect to the full potential of informative assessment teachers can:

1. Clearly share objectives and criteria for success with students.
2. Lead classroom discussions, questions, and learning activities.
3. Activate students by having them assess their own work.
4. Provide students with feedback about what they need to improve.
5. Motivate students to be instructional resources for each other (William, 2008).

MAKE AN ACTION PLAN

Teachers can use good questions to put their own action plans in place. Each teacher needs to make a specific plan about what he or she wants to change. Have teachers voice their expectations. Teachers who concentrate on making a small number of changes and really integrate them into practice make more progress. It is important for teachers to identify how they are going to make time for the new strategies listed above. Some teachers may want to revise their action plans. It is important to provide time for teachers to think it through and check to see if they made progress (Gareis & Grant, 2008).

Differentiated Lesson and Informative Assessment Plans Beginning Science and Math Skills

Differentiated and Informative Lesson and Assessment Plan 1

Sum Slide
(with Auste Siauryte)

Topic: Students will use a hundreds chart to find the sums.

Grade level: Grades 1 and 2.

Objectives: What do you want students to learn? Throughout the year students have mastered the first standard of being able to count until 200.

- Students have to learn to add and subtract two-digit problems with or without regrouping.
- The purpose of this lesson is to allow students opportunities to engage in hands-on learning by incorporating both concepts and showing them ways they can use the hundreds chart.

Why are the concepts important?
This lesson combines core first and second grade standards where students are asked to demonstrate their knowledge of number values up to 100 and solve problems up to two-digit addition. This lesson will integrate and allow students to practice both concepts, as well as learn yet another strategy of how a hundreds chart can be used.

What background information do students need before starting?
One-to-one correspondence, addition, ability to count to 100.

Organization and procedures: The teacher will show the hundreds chart and will ask students to count by fives until they reach a hundred (verbal linguistic).

The teacher will point to the numbers as they are saying them orally (visual/spatial). Then, students will be asked to skip count by fives using their own hundreds chart.

Students will be shown an example of how to find a sum of 38 + 47 = 85.

At first, students are asked to solve 38 + 47, using conventional methods and later, students are shown how to solve the same problem using the hundreds chart. Students will be given manipulatives and a hundreds board, and will be asked to practice five problems using the hundreds chart.

Materials: Hundreds chart, a pencil, manipulatives.

Lesson development, questions, and desired product: First students are taken through a process of recognizing that a hundreds chart can be used to solve addition problems.

After modeling a few problems, students will be asked to try to solve a few of their own using manipulatives. After allowing students plenty of time practicing given problems, students will be asked to pick two partners. One partner will be using the hundreds chart and the other partner will be using the conventional method. The activity will be stopped after the first partner finishes. The goal of this "competitive" method is to see which technique yields the most accurate and earliest results.

Then, students will be asked to provide pros and cons useful to their own learning and their evaluation about whether or not they found this technique useful.

Content standards addressed in this lesson:

Number sense

- Students understand and use numbers up to 100.
- Students demonstrate the meaning of addition and subtraction and use these operations to solve problems.
- Students solve addition and subtraction problems with one- and two-digit numbers.

Algebra and functions: Students use number sentences with operational symbols and expressions to solve problems.

Students write and solve number sentences from problem situations that express relationships involving addition and subtraction.

Specific differentiation: Students will be shown a model that could help with the activity. Students who still struggle with one-to-one correspondence (i.e., those who are not able to count to 100) will be asked to count using the hundreds chart, moving one number at a time; students will be asked to skip count by 2s, 5s, and 10s.

Hundreds Chart
1 2 3 4 5 6 7 8 9 10
11 12 13 14 15 16 17 18 19 20
21 22 23 24 25 26 27 28 29 30

31 32 33 34 35 36 37 38 39 40
41 42 43 44 45 46 47 48 49 50
51 52 53 54 55 56 57 58 59 60
61 62 53 64 65 66 67 68 69 70
71 72 73 74 75 76 77 78 79 80
81 82 83 84 85 86 87 88 89 90
91 92 93 94 95 96 97 98 99 100

Remember sliding down one space is the same as moving forward ten spaces

24 + 15 =	25 + 27 =	45 + 22 =	57 + 23 =
11 + 17 =	44 + 38 =	68 + 31 =	76 + 13 =
33 + 54 =	49 + 40 =	30 + 43 =	81 + 16 =

Differentiated Lesson and Informative Assessment Plan 2

Exploring Multiplication
(*with Katrina Finley, Marissa Lai, and Jennifer Zubia*)

Grade level: Grade 3.

What do you want students to learn?

- Students will multiply numbers.
- Students will solve problems by using manipulatives.

Science and math standards: Students will calculate and solve problems.

Entry skills: Counting, addition, grouping, multiplication facts.

Science and math content: Investigating the multiplication tables between 1 and 10.

Materials: Paper, stickers, chalkboard.

Modeling:

1. Ask two students to come to the front.
2. Ask the class how many eyes does each person have?
3. Ask how many eyes in total?
4. Have students describe how they got the answer. (2 + 2)
5. What is another way to get the answer the multiplication way?

    2 people times 2 eyes = 4. Repeat the process with 3 people.

Guided practice:

1. Students will be instructed to fold papers twice: one longways (hot dog style) and one shortways (hamburger style).

2. Then, they will copy the problems coordinating boxes from the board.
    1. 4 × 3
    2. 3 × 4
    3. 2 × 5
    4. 5 × 2
3. Explain to students that the first number of the problem stands for how many groups there are and the second number is how many students are in each group.
4. In 4 × 3 there are four boxes. Students are instructed to place the appropriate number of stickers on each box. (They should put three stickers in each box.) The teacher models this on the board.
5. A tiered challenge approach can be used here. All students make a problems box. For beginners, a problems box could display how many groups there are and list how many stickers each box holds. More advanced learners can compare and contrast the different amount of stickers for each group. Question: How did you decide which box a particular sticker fits into?

Evaluate your selections. As a final activity, have students create a problem box in a new and different way.

Independent practice: Students will make up their own problems to solve. The teacher will observe, make suggestions, and help as needed.

Informative assessment: Students will multiply numbers and solve problems. The teacher will assess students by completion of individual/group work.

1. Teachers explain lesson objectives and how students can be successful.
2. Teachers discuss multiplication, ask questions, and describe multiplication activities.
3. Have students go over their own work.
4. Provide individual students with feedback about what they need to improve on.
5. Encourage students to help and be resources for each other.

Closure: Students will share some of their multiplication problems. Through their examples, the teacher can reemphasize multiplication principles.

Accommodations: During "Independent Practice," make sure students understood the directions. Go over any vocabulary with them and work with students on a few problems.

If making up the problems is too complex, write out problems for them to work on. Check in on their progress periodically.

Measurement

Differentiated Lesson and Informative Assessment Plan 3

Tree Measurement
(with Doreen Toofer)

Grade level: Grade 4.

Topic: Science/math measurement.

What do you want students to learn?
This activity is used to help students understand vertical and horizontal measurement of large objects.

Objectives:

Students will be able to:

1. Demonstrate measurement of the trunk, crown, and height using vertical and horizontal measurements.
2. Compare their results with those of other groups.
3. Create a graph of their findings for the trunk, crown, and height.

Materials: String, ruler, paper, pencil, meter stick, tree, portfolio

Standards: Inquiry, life science, science and technology, personal perspectives, written communication.

Procedures:

• Students will work with a partner and go outside to find a tree to measure.
• Students will measure the tree's trunk, crown, and height as much as possible.
• Have groups compare answers and then, remeasure the tree as needed. Usually, it takes several measurements.

Trunk:

1. Measure from the ground to 4.5 feet high on the trunk.
2. At that height, measure the trunk's circumference. Use a string to measure around the trunk and record the length of the string.
3. Round to the nearest inch. Record the number and label as circumference.

Crown:

1. Find the tree's five longest branches.
2. Put markers on the ground beneath the tip of the longest branch.

3. Find a branch that is opposite it and mark its tip on the ground.
4. Measure along the ground from the first marker to the second marker.
5. Record the number and label as crown.

Height:

1. Have your partner stand at the base of the tree.
2. Back away from the tree, holding your ruler in front of you in a vertical position. Keep your arm straight. Stop when the tree and the ruler appear to be the same size. (Close one eye to help you line it up.)
3. Turn your wrist so that the ruler looks level to the ground and is horizontal. Keep your arm straight.
4. Have your partner walk to the spot that you see as the top of the ruler. Be sure the base of the ruler is kept at the base of the tree.
5. Measure how many feet he or she walked. That is the tree's height.
6. Round to the nearest foot and record your answer as the height.

Follow-up activity:

• Have students make bar graphs using information gathered outside. Have students locate the biggest and smallest trees of the same species.
• For the bar graph, students will be given markers and chart paper to create a bar graph from the information they gathered.

Formative assessment: Teachers will assist students in their measurements and provide feedback. Students should be able to correctly write down their measurements and label them. Teachers will go over student worksheets with all their data, making sure that students are measuring length, width, and height correctly. In addition, the bar graph will be a good assessment of their findings and data organization. Students will reflect on their lesson and put their data and measurement results in their portfolio.

STEM Lessons and Informative Assessment

Measuring Sticks

Process skills: Observing, inferring, communicating (sharing), and comparing.

Description: Oftentimes, elementary and middle school students are quick to compare and criticize others. This activity stresses similarities and attempts to play down differences. The method used in this activity is creative drama. Creative dramatics involves children in group interaction. It is especially useful for reaching out to ESL (English as a second language) learners, allowing them to work with other students while developing oral language. Creative dramatics is a drama for the actors themselves, connecting their feelings and attitudes with reading, literature, and other language skills.

In recent years, teachers have become interested in finding ways to bring students' values and moral decisions into their classrooms. Activities drawn from these areas involve and motivate the youth partly because they focus on issues that students really care about. Students are particularly stimulated by questions that deal with their own values, such as: Who am I? What do I care about? How am I perceived by others? How might I change? What kinds of things do other people think important?

Informative assessment: Encourage students to reflect on this activity by putting their ideas in writing.

Some suggestions:

Working in groups, choose one assessment activity:

- Write an article for your class newspaper.
- Write a letter to students in other classes, suggesting issues for debate.
- Make a video of the outstanding debate in your class.
- Reflect on the lesson.
- Put your work in your portfolio.
- Have students provide feedback to other groups, including suggestions.

Life Science

A STEM Lesson and Informative Assessment Plan 5

Leaf Exchange

The falling motions of a leaf are very complex.
Sometimes the leaf may drift randomly to the right or left as it falls.
At other times, it may tumble erratically while maintaining a downward course
… suggesting its motion is chaotic.

—Peterson, 1994

With the dramatically colorful fall season, it seems like an ideal time to set up a leaf exchange with students in other regions of the country. If it's not fall—or everything stays green all year—the shapes of leaves can also be quite interesting. Students are naturally captivated with gathering, comparing, and preserving leaves. This activity gets students involved in collecting and gathering many different types of leaves; they examine, sort, classify, investigate, and discover patterns.

Class-to-class leaf exchange is similar to writing to pen pals. Leaf exchanges get everyone interested and active—from collecting and classifying leaves into a range of sizes, shapes, and textures, to observing similarities and differences in leaves. Students learn to appreciate the value of exchanging data and gain experience in communicating with students in another region of the country. Student awareness is heightened to the wide variety of plants and trees in their own environment. A class map is then set up to track the journeys of their leaves.

Process skills: Gathering, identifying data, communicating, exploring preserving methods, comparing, graphing, and sharing.

Planning group: Members should arrange the classroom and materials.

Description: Communicating at a personal level for elementary and middle school students can be very rewarding and exciting.

Objectives:

1. Students will collect and gather leaves.
2. Students will choose a leaf exchange partner.
3. Students will sort, classify, and discover patterns.
4. Students will participate in preserving the leaves they gathered.
5. Students will improve their observation skills and increases their awareness of the plants in their environment.
6. Students as leaf pals will learn how their environment compares with those of others.
7. Students will learn to appreciate the value of exchanging data.

Standards: Inquiry, life science, personal social perspectives, written communication.

Procedures:

1. Introduce students to leaf exchange, a class-to-class leaf exchange is similar to having pen pals. Explain they are going to take part as a class exchanging preserved leaves with students in another area of the country.
2. The first step in this project is to gather leaves for the exchange. Have students bring in leaves from their neighborhoods. Or, take students on a nature walk.
3. Students are to collect at least ten different leaves.
4. When returning to the classroom, focus students' attention on the leaves' colors, shapes, and sizes.
5. Most students will want to identify the leaves. Using reference materials (slides, computer disks, prints, etc.) have students label each leaf.

Methods for preserving the leaves:

1. *Press/dry method*—place and press the leaves in an old telephone book, newspaper, or magazines; apply bricks or weights on top (to permanently preserve use a laminating process)
2. *Quick iron method*—preheat iron to permanent press setting. Place a thin sheet of cardboard on the ironing board.
   - Cut two pieces of wax paper that are larger than the leaf to be pressed. Lay one piece of wax paper on top of the cardboard, place the leaf on top of the wax paper, and put the second piece of wax paper on top of the leaf.
   - Cover the wax paper with a cotton rag, and then, press with the iron, each part of the leaf for at least twenty seconds.
   - Remove the rag, cool the leaf for about two minutes, and carefully peel the wax paper from the leaf.
3. *Preserving leaves with glycerin*—materials: 1.5 L water, 750 mL glycerin (available at drug store), a 500-mL measuring cup, newspaper, a shallow pan (You will have enough room in the pan to preserve about seventy leaves. The remaining solution can be saved.)
   - Mix the glycerin and water in the pan.
   - Place the leaves in the solution, making sure that each leaf is completely coated, and that the students are wearing aprons and goggles.
   - Soak the leaves in the solution for twenty-four hours.
   - The next day, remove the leaves and press each one between newspapers for three days or until completely dry.
   - These pressed leaves will be flattened making them easier to mail.
4. *Lamination*—If teachers have access to laminating equipment, they can preserve leaves for many years.
   - Dry and press leaves using one of the drying methods, make sure leaves are dry.
   - Follow laminating machine directions.
   - Insert the leaves like a flat piece of paper.
   - Trim the lamination film around the leaf.

Before packaging the leaves, have students compose a letter to their leaf partner describing the leaves and why they chose to send them.

Informative assessment: Have students make a wall map charting where their leaves were sent. Research the vegetation of the areas, and do a comparison analysis. Store the data in a computer. Have students reflect on the lesson and put their findings in their portfolio. Have students provide feedback to each other.

Physical Science

A STEM Lesson and Informative Assessment Plan 6

Exploring Supports

Description: There are many kinds of structures that can be described in the natural world. This beginning activity attempts to show how different kinds of structures are related. In this activity, children will find out about supports. The skills introduced include experimenting, testing strength and durability, comparing size and weight, recording data, and communicating. Introduce the activity by talking about structures, the classroom, and tables in the room. Generate questions such as, "What makes the ceiling stay where it is?", "What keeps the table from falling?" Students will soon come up with the idea of structural support. The walls of the classroom and the ceiling supports hold the ceiling up and the legs on the table keep it from falling. Discuss these ideas with the class. Explain that the name we give to these items is structural support.

Process skills: Measuring, comparing, inferring, ordering by distance, formulating conclusions.

Planning group: Members should arrange the classroom and materials.

Materials: Give each of the students some items to form a tower (e.g., a cardboard box, a tall block of wood, a paper towel roll). Provide them with the following support materials: styrofoam, wood, cartons, slitted cardboard boxes, and so on. Supply some clay, sand, and white glue. Include the following art materials: paints, brushes, construction paper, scissors, paste, and felt pens.

Objectives:

1. When presented with a problem of how to support their tower, students will explore with materials.
2. Students will discuss and share their discoveries with other class members after experimenting and trying many different support structures.
3. Students will compare their tower supports with those of the other children.
4. Students will test the strength of their tower.
5. Students will modify their support structure by adding a balcony.
6. Children will decorate their towers.
7. Students will present their investigation by answering these questions:
   - Describe how your tower and balcony are supported.
   - Show how much weight your balcony holds.
   - Explain how you made your supports.

Standards: inquiry, physical science, personal and social perspectives, written communications.

Procedures:

1. Challenge the students to find a way to make their tower stand up so that it cannot be blown over by a strong wind.
2. Helpful ideas for getting started:
   - Glue supports around the base of the tower.
   - Fill a box with sand.
   - Attach the base to a larger surface.
   - Set the tower in sand or clay.
3. After students have determined a way to support their towers, have them share what they found out.
4. Children can compare their solutions and test their towers to see how strong they really are. For example, students may decide to test their tower by having six or more students blow on it at once. Or, they could place a fan near their tower to see if it continues to stand up.
5. Encourage students to experiment further with different supports to make their towers as sturdy as they can be.
6. Have children decorate their towers with the art supplies provided.

Informative assessment: To find out how much they learned about supports, present students with another challenge. Using any of the materials, have them construct a balcony. Then, have them test the balcony by adding weights. How many weights (if any) will their balcony hold? Add more and more weights until the balcony begins to show signs of collapsing.

(This activity was adapted from Joan Westley's *Constructions*, p.22.)

A STEM Lesson and an Informative Assessment Plan 7

Interdisciplinary Bridge Building

This is an interdisciplinary activity which reinforces skills of communication, group process, social studies, language arts, mathematics, science, and technology.

Purpose and objectives: This is an interdisciplinary science and math activity which reinforces skills of communication, group process, social studies, language arts, mathematics, science, and technology.

Standards: Students use the science and math standards of collaborative learning, investigation, experimentation, measurement, and reasoning.

Materials: Lots of newspaper and masking tape; one large, heavy rock; one cardboard box. Have students bring in stacks of newspaper. You need approximately one foot of newspaper per person. Bridges are a tribute to technological efforts which employ community planning, engineering efficiency, mathematical precision, aesthetics, group effort, and construction expertise.

Procedures:
1. For the first part of this activity, divide students into three groups. Each group will be responsible for investigating one aspect of bridge building.

Group One: Research
This group is responsible for going to the library and looking up facts about bridges, collecting pictures of kinds of bridges, and bringing back information to be shared with the class.

Group Two: Aesthetics, Art, Literature
This group must discover songs, books about bridges, paintings, artwork, etc., which deal with bridges.

Group Three: Measurement, Engineering
This group must discover design techniques, blueprints, angles, and measurements of actual bridge designs. If possible, visit a local bridge to look at the structural design, etc. Pictures also help.

2. Each group presents their findings to the class.The second part of this activity involves actual student bridge construction.
   a. Assemble the collected stacks of newspaper, tape, the rock, and the box at the front of the room. Divide the class into groups of four or five students.
   b. Each group is instructed to take an even portion of newspapers to their group and one or two rolls of masking tape. Explain that the group will be responsible for building a stand-alone bridge using only the newspapers and masking tape. The bridge is to be constructed so that it will support the large rock and so that the box can pass underneath.
   c. Each group is given three to five minutes of planning time in which they are allowed to talk and plan together. During the planning time, they are not allowed to touch the newspapers and tape, but they are encouraged to pick up the rock and make estimates of how high the box is.
   d. At the end of the planning time, students are given ten to twelve minutes to build their bridge. During this time, there is no talking among the group members. They may not handle the rock or the box, but only the newspapers and tape. (A few more minutes may be necessary to ensure that all groups have a chance of finishing their constructions.)

Informative assessment: Stop all groups after the allotted time. Survey the bridges with the class and allow each group to try to pass the two tests for their bridge. (Does the bridge support the rock and does the box fit underneath?) Discuss the design of each bridge and how they compare to the bridges researched earlier.

Feedback: Enrichment: As a follow-up activity, have each group measure their bridge and design a blueprint, (include angles, length, and width of the bridge) so that another group could build the bridge by following this model.

Differentiated Lesson and Informative Assessment Plan 8

Building Visual Models: Concept Circles

In teaching the STEM subjects, teachers can make use of a variety of diagrams to help students grasp important concepts. Like mapping, concept circles demonstrate meaning and develop visual thinking. Have students represent their understanding of science concepts by constructing concept circles following these rules.

1. Let a circle represent any concept (plant, weather, bird, etc.).
2. Print the name of that concept inside the circle.
3. When you want to show that one concept is included within another concept, draw a smaller circle within a larger circle—for example, large circle for planets, smaller circle for the earth.
4. To demonstrate that some elements of one concept are part of another concept, draw partially overlapping circles. Label each (water contains some minerals). The relative size of the circles can show the level of specificity for each concept. Bigger circles can be used for more general concepts, or used to represent relative amounts.
5. To show two concepts are not related, draw two separate nonconnected circles and label each one (bryophytes—mosses, without true leaves; tracheophytes—vascular plants with leaves, stems, and roots).

Objectives:

1. Students will work with a partner.
2. Students will choose topic and make a concept circle.
3. Students will describe their visual model.

Formative assessment: Students will explain how the concepts are related. Students will write about how two concepts are not related and explain their chart. Students will put their written reports and charts in their portfolios. Have students provide feedback to each other.

Innovative Interdisciplinary Activities

Differentiated Lesson and Informative Assessment Plan 9

Active Play in Art and STEM

Active play in Art and STEM is important not only in the development of intelligence of children, but it also emerges over and over as an important step in invention and discovery. Curiosity, play, and following hunches are particularly important in developing one of the most valuable scientific tools—intuition.

Objectives:

1. Ask students to bring in materials which are cheap, durable, and safe (toys, household objects, etc.). With older students, you may want to include hammers, nails, bolts, lumber, etc.
2. Divide students into groups of five or six. By playing with the assembled objects, have students make discoveries about the sound potential of the objects they have brought in.

Procedures: Using the objects available to their group, students are to design a device that makes sound. Encourage students to use a variety of objects in as many different ways as possible. When individual designs are complete, have students share their ideas with the group.

Informative assessment: The group must pick one design and work on its construction. The important thing during this noisy period of play is to explore the rich realm of possibilities before arriving at one solution. Instruct students to put their construction in their portfolio. Have students write about their sound solution. Encourage students to provide feedback to other groups.

Awareness of the Environment

Differentiated Lesson and Formative Assessment Plan 10

Creating a Sound Garden

Objectives: Ask students to imagine things they could hear in a garden—the sound of birds, the wind, leaves, human noises, etc. Discuss various ways sound could be generated in a garden—through wind, by walking, or sprinklers going, for example. Next, have students brainstorm ideas for a sound garden. Encourage creativity, fun ideas, and original inventions.

Materials: Suggest inexpensive things that will hold up outdoors and will be safe for other children to play with. (Things that make sounds when walked on, when touched, etc.)

Procedures: Have students draw up a design for their planned construction. Then bring ideas and items from home to contribute to the project. As a group, have them plan and construct their sound garden based on the items. Designs may have to be altered as the class progresses on the project, enabling students to see the ongoing development process as ideas progress and are adjusted to fit the materials and needs of construction.

Informative assessment: Students must determine the best way to display their sound structures so that they will be accessible to others, function well, and create the best visual display in the garden. Have students write a step-by-step plan of to design their sound garden and put their plan in their portfolio.

A sample short portfolio assessment plan:

Portfolio topic: _____
Student(s): _____
Teacher: _____ Date: _____

1. Topics, questions, procedures, reasoning, or process skills explored: ____
   _____

2. Areas of *growth* and difficulties in understanding: _____
   _____

3. Areas where the work is unfinished or could be improved: _____
   _____

4. Write three things you liked about the work and one thing you didn't: ___
   _____

5. Assessment of the following areas:
   a. Inquiry and problem-solving work: _____
   _____

   b. Reasoning, discovery, and creative thinking: _____
   _____

   c. If you consulted with a partner or a small group, how did the
      collaboration work out and how might the process be improved: _____
   _____

   d. What did you find out that was strange or new to you? _____
   _____

   e. What more would you like to know about the topic? _____
   _____

Differentiated Lesson and Formative Assessment Plan 11

Exploring Earth's Pollution

Pollution is defined as an undesirable change in the properties of the lithosphere, hydrosphere, atmosphere, or ecosphere that can have deleterious effects on humans and other organisms. A part of the task for students is to decide what an undesirable change is, or what is undesirable to them.

Tell the students they are going to classify pollution in their neighborhood and city. The classification will be based on their senses and the different spheres of the earth—lithosphere (earth's crust), hydrosphere (earth's water), atmosphere (earth's gas), and ecosphere (the spheres in which life is formed). Give each student an observation sheet or have them design an observation sheet where they could show examples of pollution for a week.
Informative assessment:

Have students reflect on their observations. Encourage students to provide feedback to each other. Instruct students to record their data and written reflections in their portfolio.

Differentiated Lesson and Formative Assessment Lesson 12

Experimenting with the Unintended Consequences of Technology:

Soap Drops Derby

Students will develop an understanding that technological solutions to problems, such as phosphate-containing detergents, have intended benefits and may have unintended consequences.

Objective: Students apply their knowledge of surface tension. This experiment shows how water acts like it has a stretchy skin because water molecules are strongly attracted to each other. Students will also be able to watch how soap molecules squeeze between the water molecules, pushing them apart and reducing the water's surface tension.

Background information: Milk, which is mostly water, has surface tension. When the surface of milk is touched with a drop of soap, the surface tension of the milk is reduced at that spot. Since the surface tension of the milk at the soapy spot is much weaker than it is in the rest of the milk, the water molecules elsewhere in the bowl pull water molecules away from the soapy spot. The movement of the food coloring reveals these currents in the milk.

Grouping: Divide class into flexible groups of four or five students.

Materials: Milk (only whole or 2 percent will work), newspapers, a shallow container, food coloring, dish washing soap, a saucer or a plastic lid, toothpicks.

Procedures:

1. Take the milk out of the refrigerator half an hour before the experiment starts.
2. Place the dish on the newspaper and pour about ½ inch of milk into the dish.
3. Let the milk sit for a minute or two.
4. Near the side of the dish, put one drop of food coloring in the milk. Place a few colored drops in a pattern around the dish. What happened?
5. Pour some dish washing soap into the plastic lid. Dip the end of the tooth-pick into the soap, and touch it to the center of the milk. What happened?
6. Dip the toothpick into the soap again, and touch it to a blob of color. What happened?
7. Rub soap over the bottom half of a food coloring bottle. Stand the bottle in the middle of the dish. What happened?
8. The colors can move for about twenty minutes when students keep dipping the toothpick into the soap and touching the colored drops.

Follow-Up evaluation: Students will discuss their findings and share their outcomes with other groups.

Formative assessment: Instruct students to reflect on the experiment. Have them describe some of the benefits of detergents and some unintended consequences. Have them research common phosphates including detergents and other cleaning products. Have students put their written report in their portfolio. Have students provide feedback to other groups with suggestions for future environmental products.

Physics

Differentiated Lesson and Formative Assessment Lesson 13

Discovering Bernoulli's Principle

STEM activities often have surprising results. For example, try a simple activity that uses a ping-pong ball and a funnel to demonstrate pressure. Show the class a ping-pong ball inside a funnel. Set up the problem. Ask how far students think the ping-pong ball will go when they blow into the funnel. What happens? Students will soon discover as they blow into the funnel, the ball doesn't move. They are proving Bernoulli's principle that when air moves faster across the top surface of a material, the pressure of the air pushing down on the top surface is smaller than the pressure of the air pushing up on the bottom surface. There are many variations of the "right answer."

Objectives:

• Present the problem to the class.
• Have students guess what is happening to the ping-pong ball.
• Encourage students to write about the experiment.

Procedures:

• Have students work with a partner.
• As a class, have them guess how the experiment works.
• Have them record their guesses in their portfolio.
• Have students do the experiment and explain what is happening.

Formative assessment: Have students explain how this activity showed the following properties of air. Air has pressure. Air has weight. Air is invisible, but it is real and it takes up space. Anything that has weight pushes or presses against things. Moving air exerts pressure. Students should be able to come up with a reasonable solution. Instruct students to provide feedback to other groups and give suggestions.

In an *in*formative assessment-centered classroom, teachers interact frequently with students on a daily basis. The teacher plays an essential role in connecting assessment to students' opportunities to understand how science takes place on the real-world. Providing students with situations where students make discoveries through their investigations often sparks new ideas and scientific ways of thinking. In the formative assessment-centered classroom, students play an active role in their own learning and support the learning of others (Keeley, 2008).

Final informative assessments: The final assessment usually occurs at the end of a unit or topic of study. It is a culminating experience involving several lessons and reflections. Following are a list of criteria for assessing lessons and unit plans:

1. State and content standards for your subject need to be listed.
2. Lesson objectives should clearly state what students will learn.
3. A rationale needs to be included explaining why it is important to teach this concept and skill at this time.
4. Prerequisite skills need to be identified.
5. It is important to look at background information before beginning the lesson.
6. Procedures should be well defined and organized.
7. Materials need to be accessible for teaching the lesson.

8. Small group options and adaptations need to be provided.
9. Techniques for gearing up the lesson if material is too easy and gearing down the lesson if too difficult need to be available to the teacher.
10. Informative assessment is an ongoing piece of a good unit or lesson.

Whatever type of lesson plan and related assessment technique you decide to use, teacher-directed experiences, small group activities, individual responsibilities, and informative evaluation techniques can all be briefly spelled out in a one- or two-page plan. (Of course, there are always factors to think about that go beyond the plan.) As far as the teacher is concerned, the instructional payoff has to be worth the effort. Less is often more because asking for too much paperwork can result in teachers' mechanically meeting the requirements—or avoiding the process altogether.

The content standards for the STEM subjects have many good ideas and teaching suggestions. They have a lot to do with helping students acquire knowledge by thinking critically, inquiring collaboratively, and using an assortment of intellectual tools to gain knowledge about the world. It also means making sure that students get a sense of the subject and the connections between subjects.

Assessment today is much more than teaching a chapter or unit and giving a test at the end. There needs to be intermediate checks (formative assessment) of understanding if learning is going to be maximized. Whether you call it formative, informative, or transformative assessment, student/teacher feedback, discussion and observation during class activities provide useful information about learning while it is happening. The next step is quickly acting on that information in a way that improves student learning.

When done right, ongoing assessment stimulates students and encourages them to assume ownership of their own learning. Along the way, they can learn to assess their own work using agreed-upon criteria for success.

Neither collaborative inquiry, problem solving, nor anything else should rule out teaching for content. In science, for example, inquiry frequently comes into contact with basic scientific principles. Another example is mathematics where interpretation and discovery don't rule out calculation; reasoning about numbers to *understand* a problem is not antithetical to reasoning with numbers to *get* an answer.

SUMMARY, CONCLUSION, AND LOOKING AHEAD

A STEM lesson plan is a format for implementing the teacher's goals and objectives in a way that connects to the district curriculum and subject matter standards. It includes activities and methods of assessment. Experienced teachers recognize the fact that there should be space for reasoning skills and the creative imagination in any lesson plan. It is possible to provide everyone in the classroom with the powerful intellectual tools they need to imagine new approaches and unique applications.

Some of our students excel at seeing the beauty of things; others are better at seeing things and wanting to make them better. Sometimes the two things balance out. So, it is important to make the classroom a place where it is possible to find fresh ideas, appreciate them, and come up with unique and unexpected answers.

In the classroom, the teacher can focus on giving learners the opportunity to use their ideas and the information they gather to inquire about a topic and propose unique solutions to problems. In such a supportive learning community, time is set aside for personal reflection and engaging feedback from peers. It is also important to note that thinking and interaction do not stop when the inquiry is over or the problem has been solved.

It is clear that competency in the STEM subjects is an important building block in the foundation of academic development, life experience, and the human imagination. It involves more than being creative at a personal level. Some properties of the various subjects are wide open for discovery and debate, while other things are known and settled.

It is up to the instructor to help students recognize the difference, while arousing natural curiosity and creative thinking. At their best, schools are places where students are taught to recognize, respect, and critically examine brilliant ideas by great thinkers (Mandinach & Gummer, 2016).

Putting standard-based, research-driven instruction into practice doesn't start and stop with the teacher. It requires administrative support at all levels. It also requires supportive parents for can assist in helping learners to take responsibility for developing their own understanding.

Directly or indirectly, everyone is involved in the education of children and young adults. When it comes to informed citizenship, we must all have some understanding of the role of science and math in our lives today. Just as importantly, we all need to consider how the products of both domains will become even more central to our lives in the future.

If the STEM subjects are taught well, students will find them useful, interesting, and even beautiful. In school, as in life, it is important to have the space needed to make mistakes. And it is just as important to develop the skills needed, learn from mistakes.

The future is not some place we are going to, but one we are creating. The paths to it are not found but made, and the activity of making them changes both the maker and the destination.

—John Schaar

REFERENCES

Chapman, C. & King, R. (2008). *Differentiated instructional management: Work smarter, not harder*. Thousand Oaks, CA: Corwin Press.

Gardner, H. (2006). *Five minds for the future*. Boston, MA: Harvard Business School Press.

Gareis. C. & Grant, L. (2008). *Teacher-made assessments: How to connect curriculum, instruction, and student learning*. Larchmont, NY: Eye On Education.

Mandinach, E. & Gummer, E. (2016). *Data literacy for educators*. New York, NY: Teachers College Press.

Peterson, I. (1994). "Catching the flutter of a falling leaf." In *Science news* J. Miller and B. Potter, eds. Washington, DC: Science Service Publication.

Popham, W. J., (2008). *Classroom assessment: What teachers need to know* (5th edition). Boston, MA: Allyn and Bacon.

Keeley, P. (2008). *Science formative assessment: 75 practical strategies for linking assessment, instruction, and learning*. Thousand Oaks, CA: Corwin Press.

Schaar, J. (2016). GoodReads web site, "Get Quotes Daily." From link: http://www.goodreads.com/quotes/279924-the-future-is-not-some-place-we-are-going-but

Scherer, M. (2016). *On formative assessment*. Alexandria, VA: Association for Curriculum and Curriculum Development.

Stiggins, R., Arter, J., Chappuis, J., & Chappuis, S. (2006). *Classroom assessment for student learning: Doing it right—using it well*. Portland, OR: Educational Testing Service.

Sousa, D. (2008). *How the brain learns mathematics*. Thousand Oaks, CA: Corwin Press.

Tomlinson, C., Brimijoin, K., & Narvaez, L., (2008). *The differentiated school: Making revolutionary change in teaching and learning*. Alexandria, VA: Association for Supervision and Curriculum Development.

William, D. (2008). Changing instructional practice. *Educational Leadership* 65:4, 36–41.

About the Authors

Dennis Adams is a former elementary school teacher who has taught at the University of Minnesota, University of Maine, and McGill University in Montreal. He holds a Ph.D. from the University of Wisconsin and did post-doctoral work at Harvard University. He is the author of more than twenty-five books and more than a hundred journal articles on various educational topics.

Mary Hamm has taught at Ohio State University and the University of Colorado. More recently, she has been teaching at San Francisco State University. She has worked on both math and science standards and has published more than a dozen books and eighty journal articles on these subjects.

CPSIA information can be obtained
at www.ICGtesting.com
Printed in the USA
BVOW08s2357240217
477150BV00001B/1/P